Joint RES and Distribution Network Expansion Planning Under a Demand Response Framework

Joint RES and Distribution Network Expansion Planning Under a Demand Response Framework

Javier Contreras, Miguel Asensio, Pilar Meneses de Quevedo, Gregorio Muñoz-Delgado, and Sergio Montoya-Bueno
School of Industrial Engineering, University of Castilla - La Mancha, Ciudad Real, Spain

ELSEVIER

AMSTERDAM • BOSTON • HEIDELBERG • LONDON
NEW YORK • OXFORD • PARIS • SAN DIEGO
SAN FRANCISCO • SINGAPORE • SYDNEY • TOKYO

Academic Press is an imprint of Elsevier

Academic Press is an imprint of Elsevier
125 London Wall, London EC2Y 5AS, UK
525 B Street, Suite 1800, San Diego, CA 92101-4495, USA
50 Hampshire Street, 5th Floor, Cambridge, MA 02139, USA
The Boulevard, Langford Lane, Kidlington, Oxford OX5 1GB, UK

British Library Cataloguing-in-Publication Data
A catalogue record for this book is available from the British Library

Library of Congress Cataloging-in-Publication Data
A catalog record for this book is available from the Library of Congress

ISBN: 978-0-12-805322-5

For Information on all Academic Press publications
visit our website at http://www.elsevier.com/

  Working together
to grow libraries in
developing countries

www.elsevier.com • www.bookaid.org

Publisher: Joe Hayton
Acquisition Editor: Lisa Reading
Editorial Project Manager: Maria Convey
Production Project Manager: Anusha Sambamoorthy
Cover Designer: MPS

CONTENTS

CHAPTER 1

Introduction

An energy system usually consists of generation units, transmission networks, distribution networks, consumption centers, and control, protection, and regulation equipment [1]. Electric power systems have developed along more or less the same lines in all countries, converging toward a very similar structure and configuration. Electric power systems are conditioned by the fact that generation and demand must be in instantaneous and permanent balance. Distribution networks are an important part of the electrical system, since they supply energy from distribution substations to end users. Distribution networks are typically three-phased, and the standard operating voltages are 30, 20, 15, and 10 kV. The structure of this medium-voltage network may vary, but its operation is always radial. Substations normally house circuit-breakers that protect the so-called feeders, ie, lines running to other transformer stations where the voltage is stepped down again to supply low-voltage power. These networks have been designed with wide operating ranges, which allows them to be passively operated resulting in a more economical management.

Distribution substations are fed through one or several medium-voltage networks, although sometimes, can be directly connected to high-voltage networks. Each distribution substation meets the energy by means of one or several primary feeders. Generally, a distribution substation contains: (1) protection devices, (2) measure devices, (3) voltage regulators, and (4) transformers [2].

From a centralized standpoint, distribution companies are responsible for operation and planning. Distribution companies must satisfy the growing demand with quality and in a secure fashion. Therefore, planning models are used to obtain an optimal investment plan at minimum cost meeting the security and quality imposed requirements. These planning models are based on capacity distribution network expansion considering: (1) replacement and addition of feeders, (2) reinforcement of existing substations and the construction of new substations, and (3) installation of new transformers [3].

Diesel and heavy fuel oil generation units currently dominate the generation mix in small islands. The natural increasing load factors in peak periods are putting stress on real systems, leading to the use of peak load plants to cover those consumption peaks and, therefore, incurring higher costs for the whole system. Islands will face in the future considerable challenges in order to meet their energy needs in a sustainable, affordable, and reliable way.

The main reasons behind this choice are the relative ease to purchase fuel (compared to liquefied natural gas or compressed natural gas), the flexibility of the installed engines in meeting daily and seasonal variations in energy demand, and the lack of storage. Furthermore, the need to maintain adequate levels of redundancy in case of a failure of a producing unit has led companies to prefer several smaller units instead of one large generator. In addition, diesel engines' efficient operation across volatile demand scenarios, along with their relatively low installation and maintenance costs, has made this technology the backbone of most island power generation systems.

Policy makers have promoted wind and solar generation in an effort to improve the sustainability of electric power systems, being subject to different incentive schemes that have led to a significant penetration of those technologies. Isolated energy systems share common characteristics and are subject to common challenges, especially from the security of supply and system stability perspectives. Threats experienced by isolated systems as a consequence of increasing DG penetration are even higher since they cannot depend on the smoothing effect of a large balancing area.

Renewable generation technologies have been promoted by policy makers throughout the years in an effort to increase the sustainability of electric power systems. However, threats experienced by isolated systems as a consequence of the increasing distributed generation (DG) penetration are higher than those experienced by interconnected systems, since they cannot depend on the smoothing effect of a large balancing area and interconnection flows. In addition, renewable technologies are becoming price-competitive, especially in isolated systems, where diesel and heavy fuel oil generation units dominate the generation mix. The widespread growth of DG, mainly due to the penetration of renewable energy, inevitably requires the inclusion of this kind of generation in planning models [4–6]. In Ref. [7] the multistage expansion planning problem of a distribution system where investments in the distribution network and in the DG are jointly considered is presented.

Uncertainty is present in load demand, wind, and photovoltaic (PV) energy production. On one hand, many tools have been developed for load forecasting with good results. On the other hand, wind and PV energy production uncertainty are the most difficult to manage because of their high variability which depends on weather, where many variables are present. The increasing penetration of nondispatchable technologies in overall energy mix calls for the inclusion of uncertainty associated to production stochasticity of renewable technologies (wind, solar) involved in the expansion planning model.

An algorithm based on Refs. [3] and [8] is developed to decide the joint expansion planning of DG and distribution network. The optimization model decides the addition, replacement, or reinforcement of different assets such as feeders, transformers, substations, and generators. Planning models have been used for many years to optimize generation investments in electric power systems. However, these models have not been completely adapted in order to treat DR on an equal footing. This chapter stresses the importance of integrating DR to time-varying prices (real-time prices) into those investment models. Own-and cross-price elasticities are included in order to incorporate consumers' willingness to adjust the demand profile in response to price changes. The concept of "flexible demand" has generated significant interest but is still in the early developmental stages. Additionally, a few demonstration projects have tested the actual capability of loads to be controlled in order to cope with the challenges posed by increasing renewables penetration.

Variability and uncertainty are not unique to stochastic generation resources. Similar challenges are posed by aggregated electricity demand and, to a certain extent, by supply resources. Over the years, different techniques for managing the variability of demand and generation of the system through the use of reserves have been developed by grid operators.

The role of demand has recently attracted an increasing interest in power systems. Given the desired integration of uncontrollable renewable generation, the traditional paradigm of controllability provision only by the generation side will not be economically sustainable. Furthermore, environmental and security of supply concerns have also flatten the way for the electrification of transport and heat sectors, which is expected to introduce a significant amount of new demand in power systems.

Additionally the integration of energy storage systems (ESS) in distribution networks may lead to challenges coming from the intermittency and uncertainty of renewable energy resources (RES) and the economic aspects of the integration of ESS in electrical distribution systems. ESS may contribute to increase the integration of the power generated by RES [9] and also to reduce generation costs and improve the quality of the power supply [10]. Nevertheless, ESS are still expensive [11, 12].

1.1 HISTORICAL BACKGROUND AND MOTIVATION

Planning models have been used for many years to optimize generation investments in electric power systems. However, these models have not been completely adapted in order to treat price-dependent resources, such as DR or ESS, on an equal footing.

Distribution system planning strategies have traditionally followed an established rule-based experience. A new capacity has been built once the load growth value has reached a predetermined threshold. This new capacity has been obtained by expanding the capacity of pre-existing substations or considering the addition of new substations. Authors in Refs. [13−18] propose different approaches to solve a Distribution Company (DISCO)'s distribution expansion planning. Under this framework, DISCOs aim to improve their profits and minimize the investment risk to meet the growth demands in their territories on a reliable way while keeping the electricity bills for their consumers affordable.

Distribution networks were traditionally not designed to accommodate generation and therefore DISCOs have not included DG in their expansion planning. Renewable generation technologies have been promoted by policy makers throughout the years in an effort to increase the sustainability of electric power systems. In addition, renewable technologies are becoming price-competitive, especially in isolated systems, where diesel and heavy fuel oil generation units dominate the generation mix. This situation has completely modified the planning, operation, and maintenance of distribution networks for the DISCOs [19−32].

The role of DR and ESS has recently attracted an increasing interest in power systems. However, previous models have not been completely adapted in order to treat DR and ESS on an equal footing. Authors in Refs. [32] and [33] analyze the effects of integrating short-term DR into long-term investment planning, defining an analytical framework to incorporate DR in long-term resource planning. An hourly model is proposed, stressing the importance of integrating short-term DR to time-varying prices into long-term investment models. Own- and cross-price elasticities are used to quantify willingness of consumers to modify their consumption patterns. A network investment planning model for high penetration of wind energy under DR program is presented in Ref. [34]. The transmission planning problem is modeled in the framework of a linear optimization problem and a nonlinear optimization process is used to analyze the Monte Carlo results. In Ref. [35] the effects of DR on generation expansion planning in restructured power systems are modeled. Authors in Ref. [36] provide an alternative approach to model load levels in electric power systems with high renewable penetration. The approach introduces the concept of system states as opposed to load levels, allowing a better incorporation of chronological information in power system models.

This book enhances previous approaches considering that a symmetric treatment of load and generation creates the strongest possible incentive for final consumers to actively participate in the wholesale electricity market. This book translates this argument to medium- and long-term distribution and generation expansion planning procedures. DR and ESS will play more and more a significant role in future expansion planning model. This book stresses their relevance, including additional considerations to the planning model.

The main objective of this book is the implementation of an algorithm to decide the joint expansion planning of DG and the distribution network. The outcomes of the model are the locations and sizes of new generation assets to be installed when fixed and variable costs are known. The model introduces other issues relevant to planning in insular distribution systems, including DR and hybrid storage.

1.2 OUTLINE

This book contains eight chapters, including an introductory chapter where the addressed topic is introduced and the available technical literature is presented. Notation used in this chapter will be used throughout the book. In chapter "Renewable Power Generation Models," the renewable production models for wind and PV are described. The considered production models relate the RES (wind speed and irradiation) with the output power of each generator. In chapter, "Uncertainty Modeling," uncertainty modeling is developed taking into account the load, wind, and PV uncertainty. Motivation for the proposed approach to approximate demand curve by load levels making the problem tractable is detailed. Chapter "Demand Response Modeling" reflects the analytical framework to incorporate DR in long-term resource planning, stressing the importance of incorporating price-varying DR into investment models. Effects of DR to time-varying prices are presented. ESS are described in terms of benefits and technologies in insular grids are described in chapter "Energy Storage Systems Modeling." Additionally the generic energy storage device used in this expansion planning is presented. The mathematical formulation for the optimization problem of distribution network expansion planning is presented in chapter "Optimization Problem Formulation." A multistage planning framework is presented where radial operation of the distribution network is explicitly imposed an approximate network model is used. Costs of losses are included in the objective function and several investment alternatives exist for each asset. A case study based on a real distribution network (La Graciosa in the Canary Islands), composed of 26 nodes and 37 branches, is presented and used for demonstration in chapter "Case Study." A base case has been defined for the comparison of the results, where neither DR nor ESS is considered in generation and distribution expansion planning. The model has been enhanced, including the integration of DR in generation and distribution expansion planning, the integration of ESS in generation and distribution expansion planning, and the integration of both DR and ESS in generation and distribution expansion planning. A cost−benefit analysis of the considered case studies is provided. The last chapter includes a summary of the work is presented, the conclusions drawn, and future work is proposed.

CHAPTER 2

Renewable Power Generation Models

This chapter provides a brief overview on the generation models developed for renewable distributed generation units. These production models relate the renewable energy resource with the output power of each generator. Despite the vast range of renewable technologies, the section focuses on wind and PV technologies. Hybrid storage technologies will be discussed in the corresponding section.

2.1 PHOTOVOLTAIC ENERGY

2.1.1 Introduction

The photoelectric effect consists of converting the sun's radiation into electricity. This process is achieved by the property of some materials to absorb photons and emit electrons. The materials used in PV cells are semiconductors (see Fig. 2.1).

The equivalent circuit for a PV cell consists of a real diode in parallel with an ideal current source. The ideal current source delivers a current proportional to the solar flux or irradiance. Furthermore, the behavior is modeled by adding a resistor in parallel and in series. The equivalent circuit is shown in Fig. 2.2.

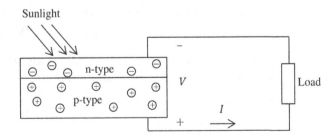

Figure 2.1 Photovoltaic cell diagram.

Joint RES and Distribution Network Expansion Planning under a Demand Response Framework.

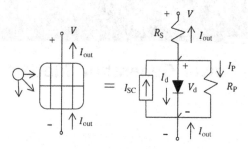

Figure 2.2 Real circuit for a PV cell.

The Kirchhoff's expression of the equivalent circuit for a PV cell results as follows:

$$I = I_{SC}C - I_0(e^{qV_d/kT} - 1) - V_d/R_p \tag{2.1}$$

where:

V_d = voltage across the diode terminals (V);
q = electron charge (1.6×10^{-19} C);
k = Boltzmann's constant (1.38×10^{-23} J/K);
T = temperature (K);
I_0 = reverse saturation current (A).
R_P = parallel resistance (Ω).

2.1.2 Power Output and *I–V* Curves

The power curve is the most important diagram of a PV module. Fig. 2.3 represents a generic $I-V$ curve and the power output for a PV module, where several parameters can be identified, such as short-circuit currents and open-circuit voltages. Additionally the diagram provides the power delivered by the module as the product of current and voltage. It can be observed how output power varies depending on current and voltage. The maximum output power (P_{max}) is reached by I_{mpp} and V_{mpp} and is the usual working point for PV panels.

The curves are affected by temperature and irradiation. The expression of the relationship between irradiation and current is:

$$I_{SC} = G\frac{I_{SC}(CEM)}{1000} \tag{2.2}$$

where:

G = solar irradiance (W/m^2);
I_{SC} = short-circuit current at CEM conditions (A).

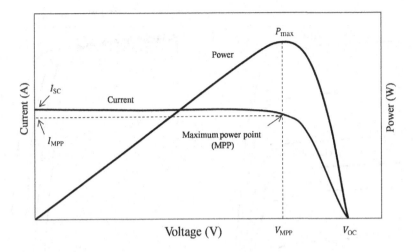

Figure 2.3 I–V curve and power output for a PV module.

The temperature of a PV cell increases with ambient temperature and reheating from irradiance. The equation that considers these effects is:

$$T_{cell} = T_{amb} + \left(\frac{NOCT - 20°C}{0.8}\right) G \qquad (2.3)$$

where:

T_{cell} = cell temperature (°C);
G = solar irradiance (kW/m²);
T_{amb} = ambient temperature (°C);
NOCT = nominal operating cell temperature conditions (°C).

Fig. 2.4 shows the effect of irradiation and temperature on I–V curves.

2.1.3 Parameters and Operating Conditions
There are some parameters and operating conditions to be defined for PV modules.

The electrical parameters are:

- Maximum power (P_{max}): corresponds to the point where the product of current and voltage is maximum. Its other name is peak power.
- Open-circuit voltage (V_{oc}): is the open-circuit voltage output (without load) of a PV cell when solar irradiance and temperature data are known.

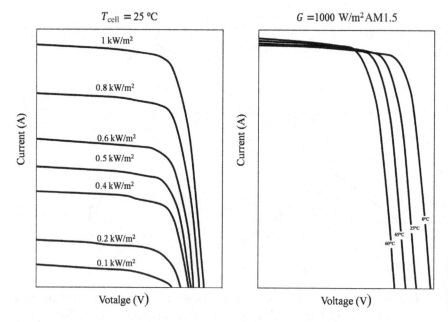

Figure 2.4 I–V curves under different temperatures and irradiance.

- Short-circuit current (I_{SC}): is the short-circuit current output of a PV cell when solar irradiance and temperature data are known.
- Voltage at the maximum power point (V_{mpp}).

The thermal parameters are:

- Current–temperature coefficient (α): indicates the variance of the short-circuit current with an increasing temperature. It can be an absolute value mA/°C or a relative value %/°C.
- Voltage–temperature coefficient (β): indicates the variance of the open-circuit voltage with an increasing temperature (negative value). It can be an absolute value mV/°C or a relative value %/°C.
- Power–temperature coefficient (δ): indicates the variance of the maximum power with increasing temperature. It can be an absolute value W/°C or a relative value %/°C.

The operating conditions are shown in Tables 2.1 and 2.2 for standard test conditions (STC) and nominal operating cell temperature (NOCT), respectively.

Table 2.1 Standard Test Conditions (STC)	
Solar irradiance	1 kW/m^2
Air mass ratio	AM1.5
Temperature	25°C
No wind	

Table 2.2 Nominal Operating Cell Temperature (NOCT)	
Solar irradiance	800 W/m^2
Wind speed	1 m/s
Temperature	20°C
Mounting = open back side	

2.1.4 Models for Photovoltaic Systems

In this section, three different models of PV production are shown.

2.1.4.1 Progensa's Model [37]

This model reflects the effect of temperature and uses the power–temperature coefficient. The output of the PV panel P_{out} is estimated using (2.4) and its efficiency can be estimated by means of (2.5).

$$P_{out} = AG\eta_{out} \qquad (2.4)$$

$$\eta_{out} = \eta_{STC}[1 + \delta(T_{cell} - 25)] \qquad (2.5)$$

where:

A = area of the PV panel (m^2);
G = solar irradiance (W/m^2);
η_{out} = efficiency in a determined operating condition;
η_{STC} = efficiency at standard test conditions;
T_{cell} = cell temperature (°C);
δ = power–temperature coefficient (%/°C).

Combining (2.4) and (2.5), the power output of a PV panel can be expressed in terms of the power obtained under standard test conditions (P_{STC}).

$$P_{out} = P_{STC}\left\{ \frac{G}{1000}[1 + \delta(T_{cell} - 25)] \right\} \qquad (2.6)$$

$$T_{cell} = T_{amb} + \left(\frac{NOCT - 20°C}{800}\right)G \qquad (2.7)$$

where:

G = solar irradiance (W/m^2);
P_{STC} = power under standard test conditions (STC) (W);
T_{amb} = ambient temperature (°C);
T_{cell} = cell temperature (°C);
δ = power−temperature coefficient (%/°C);
NOCT = nominal operating cell temperature conditions (°C).

2.1.4.2 Atwa's Model [38]

Atwa presents a simple photovoltaic panel model. The output power of the PV module is dependent on solar irradiance and ambient temperature of the site as well as the characteristics of the module itself. The cell temperature is calculated using (2.3). Current and voltage outputs for a PV panel (P_{out}) are determined by (2.8) and (2.9).

$$I_{out} = G[I_{SC} + \alpha(T_{cell} - 25)] \qquad (2.8)$$

$$V_{out} = V_{oc} + \beta T_{cell} \qquad (2.9)$$

where:

I_{out} = output current (A);
G = solar irradiance (W/m^2);
I_{SC} = short-circuit current at STC conditions (A);
α = current−temperature coefficient (A/°C);
T_{cell} = cell temperature (°C);
V_{out} = output voltage (V);
V_{oc} = open-circuit voltage at STC conditions (V);
β = voltage−temperature coefficient (V/°C).

Furthermore, the fill factor (FF) is estimated as:

$$FF = \frac{V_{mpp}I_{mpp}}{V_{oc}I_{SC}} \qquad (2.10)$$

where:

V_{mpp} = voltage at the maximum power point (V);
I_{mpp} = current at the maximum power point (A).

Then, the power output of the PV model is calculated as:

$$P_{out} = FF I_{out} V_{out} \qquad (2.11)$$

2.1.4.3 Borowy's Model [39]

Borowy proposes a model to determine the power output of a PV module. The model assumes the use of a maximum power point tracker. The maximum power point tracking (MPPT) is a technique that grid connected inverters use to obtain the maximum possible power from a PV device.

This model can be divided into several parts:

- Temperature variation
  The cell temperature is calculated using (2.3). The variation is calculated by the following equation:

$$\Delta T = T_{cell} - 25 \qquad (2.12)$$

where:
   T_{cell} = cell temperature (°C).
- Current variation
  The current depends on irradiation and temperature:

$$\Delta I = \alpha \left(\frac{G}{1000} \right) \Delta T + \left(\frac{G}{1000} - 1 \right) I_{SC} \qquad (2.13)$$

where:
   α = current−temperature coefficient (A/°C);
   I_{SC} = short-circuit current at STC conditions (A);
   G = solar irradiance (W/m^2).
- Voltage variation

$$\Delta V = -\beta \Delta T - R_s \Delta I \qquad (2.14)$$

where:
   β = voltage−temperature coefficient (V/°C);
   R_s = series resistance (Ω).
- Constants

$$C_1 = \left(1 - \frac{I_{mpp}}{I_{SC}} \right) \exp \left[\frac{-V_{mpp}}{C_2 V_{oc}} \right] \qquad (2.15)$$

$$C_2 = \frac{V_{mpp}/V_{oc} - 1}{\ln(1 - I_{mpp}/I_{SC})} \qquad (2.16)$$

where:

I_{mpp} = current at maximum power point (A);
V_{mpp} = voltage at maximum power point (V);
V_{oc} = open-circuit voltage (V).

- Module current

$$I(V) = I_{SC}\left\{ 1 - C_1\left[\exp\left(\frac{V + \Delta V}{C_2 V_{oc}}\right) - 1\right] \right\} + \Delta I \qquad (2.17)$$

where:

I = output current (A);
V = output voltage (V).

- Maximum module power output
  To determine the maximum output power of a PV model, it is necessary to evaluate expression (2.18) for different values of voltage. The voltage values are between 0 and 1.32 times V_{oc}.

$$P_{out}^{max} = VI \qquad (2.18)$$

where:
P_{out}^{max} = maximum output power (W).

2.2 WIND ENERGY

2.2.1 Introduction

Wind energy is a source of renewable power coming from the air current flowing across the earth's surface. Wind turbines harvest kinetic energy from the wind flow and convert it into usable power. The origin of wind power is shown in expression (2.19).

Considering a cube of air mass m moving at a speed v (Fig. 2.5), the kinetic energy associated is defined by the following equation:

$$E_{kinetic} = \frac{1}{2}mv^2 \qquad (2.19)$$

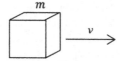

Figure 2.5 Cube of air mass m *moving at a speed* v.

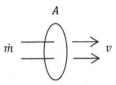

Figure 2.6 Air mass moving at a speed through area.

where:

m = mass (kg);
v = speed (m/s).

The power represented by a mass of air moving at velocity v through area A will be (Fig. 2.6):

$$P_{kinetic} = \frac{1}{2}\left(\frac{Mass}{Time}\right)v^2 \qquad (2.20)$$

The mass flow rate $\dot{m}$, through area A, can be formulated as the product of air density ρ, speed v, and cross-sectional area A:

$$\left(\frac{Mass\ through\ A}{Time}\right) = \dot{m} = \rho A v \qquad (2.21)$$

Combining (2.20) with (2.21) gives the following relation:

$$P_{wind} = \frac{1}{2}\rho A v^3 \qquad (2.22)$$

where:

ρ = air density (kg/m^3);
A = cross-sectional area (m^2);
v = wind speed normal to A (m/s).

Wind turbines turn with the moving air and power an electric generator that supplies an electric current. The wind turns the blades, spinning a shaft which is connected to a generator.

In Fig. 2.7, the upwind velocity is v, the velocity of the wind through the plane of the rotor blades is v_b and the downwind velocity

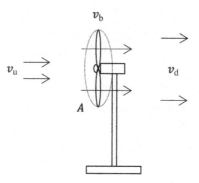

Figure 2.7 Wind through a wind turbine.

is v_d. The power produced by the blades, P_b, is equal to the difference in kinetic energy between the upwind and downwind air flows:

$$P_b = \frac{1}{2}\dot{m}(v^2 - v_d^2) \tag{2.23}$$

The mass flow rate, $\dot{m}$, of the air within the stream tube is defined as:

$$\dot{m} = \rho A v_b \tag{2.24}$$

Assuming that the velocity of the wind through the plane of the rotor is just the average of the upwind and downwind speeds, then, the new expression is:

$$P_b = \frac{1}{2}\rho A\left(\frac{v + v_d}{2}\right)(v^2 - v_d^2) \tag{2.25}$$

The ratio of downstream to upstream wind speed is defined as λ.

$$\lambda = \left(\frac{v_d}{v}\right) \tag{2.26}$$

Substituting (2.25) into (2.26) gives:

$$P_b = \frac{1}{2}\rho A\left(\frac{v + \lambda v}{2}\right)(v^2 - \lambda^2 v^2) = \frac{1}{2}\rho A v^3\left[\frac{1}{2}(1 + \lambda)(1 - \lambda^2)\right] \tag{2.27}$$

Eq. (2.27) shows how the power produced by wind is equal to the upstream power in the wind multiplied by a fraction. The fraction designated as C_p refers to the efficiency of the rotor (2.28).

$$C_p = \frac{1}{2}(1 + \lambda)(1 - \lambda^2) \qquad (2.28)$$

The efficiency of the rotor can be represented versus to the wind speed ratio.

Fig. 2.8 shows the point where the efficiency of the rotor is maximum. This point can be determined deriving (2.28) with respect to λ and set it equal to zero:

$$\frac{dC_p}{d\lambda} = \frac{1}{2}\left[(1 + \lambda)(1 - 2\lambda)(1 - \lambda^2)\right] = 0 \qquad (2.29)$$

This expression has as solution:

$$\lambda = \frac{v_d}{v} = \frac{1}{3} \qquad (2.30)$$

Substituting (2.30) into (2.28), the maximum blade efficiency is attained.

$$\text{Maximum rotor efficiency} = \frac{1}{2}\left(1 + \frac{1}{3}\right)\left(1 - \frac{1}{3^2}\right) = 59.3\% \qquad (2.31)$$

Eq. (2.31) expressed the maximum rotor efficiency that is known as Betz efficiency.

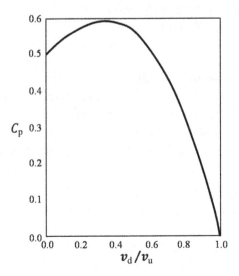

Figure 2.8 Rotor efficiency.

Table 2.3 Friction Coefficient	
Soil Characteristics	Friction Coefficient $\propto$
Smooth hard ground, calm water	0.10
Tall grass on level ground	0.15
High crops, hedges, and shrubs	0.20
Wooded countryside, many trees	0.25
Small town with trees and shrubs	0.30
Large city with tall buildings	0.40

2.2.2 Impact of Height

Wind power is proportional to the cube of wind speed. Therefore, a small variation in wind speed increases significantly the output power. The following expression is used to characterize the wind speed profile and the impact of the roughness of the earth's surface on wind speed (Table 2.3):

$$\left(\frac{v}{v_0}\right) = \left(\frac{H}{H_0}\right)^{\alpha}$$ (2.32)

where:

$\propto$ = friction coefficient;
v_0 = wind speed at height H_0 (m/s);
H_0 = reference height (m).

2.2.3 Wind Generation Model

The wind generation model converts wind speed values (m/s) into wind power (W). Turbine manufacturers provide power–wind speed data. These data can be expressed as a graph (Fig. 2.9) or as a table. Additional relevant information provided by the power–wind table is as follows:

v_{cut} = cut-in speed (m/s);
v_{rated} = rated speed (m/s);
v_{co} = cut-off speed (m/s);
P_R = rated electrical power (kW).

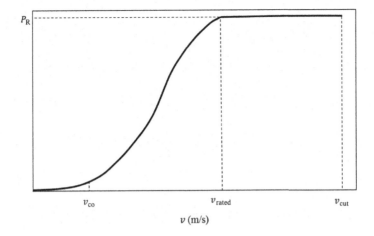

Figure 2.9 Power curve for a wind turbine.

The power output of a wind turbine is dependent on the wind speed, as already discussed. The output power along the different states is calculated using the linearized curve that is represented by the following equation:

$$
P_{out} = \begin{cases} 0, & v < v_{cut} \\ \dfrac{P_R}{v_{rated} - v_{cut}} v + P_R \left(1 - \dfrac{v_{rated}}{v_{rated} - v_{cut}} \right), & v_{cut} \leq v < v_{rated} \\ P_R, & v_{rated} \leq v < v_{co} \\ 0, & v \geq v_{co} \end{cases}
$$

$$(2.33)$$

where:

v = wind speed (m/s);
v_{cut} = cut-in speed (m/s);
v_{rated} = rated speed (m/s);
v_{co} = cut-off speed (m/s);
P_R = rated electrical power (kW);
P_{out} = output power of the wind turbine (kW).

CHAPTER 3

Uncertainty Modeling

3.1 INTRODUCTION

Generation expansion planning models have significantly evolved in the last decades, driven by the increasing penetration of renewable and nondispatchable technologies in the overall energy mix. Different effects are derived from this increasing penetration, including the need for modifying the scheduling regime and thus the unit-commitment costs of the rest of the generating facilities, increasing for instance the need of cycling conventional thermal generation. Traditionally, expansion planning models have used the screening curves approach in order to provide a way to view trade-offs between technologies in terms of investment costs and operating costs. Screening curves is a simple way of performing preliminary screening of planning alternatives, without accounting for intertemporal cost–benefit interactions and system reliability contributions of the various possible investments. The use of screening curves provides a basic tool to determine the optimal generation mix of base load, shoulder, and peaking capacities. This approach is deemed sufficiently accurate when the time profile of load does not matter in first approximation, as it is the case for dispatchable generations. However, screening curve analysis is deemed insufficient to accurately represent detailed production costs missing relevant information such as system reliability, forced outages, and renewable technologies. In this approach, load is represented by a cumulative probability distribution during a year, referred to as a "load duration curve." Demand for electricity is represented by means of a load duration curve built by sorting load levels starting from the highest level in a certain region at hourly intervals, typically for 1 year, that is, 8760 h. Under this approach, electricity demand is traditionally divided in three load periods from the longest to the shortest: base load, shoulder, and peak demand. The variability and uncertainty in renewable resources like wind and solar power pose new challenges from a long-term planning perspective. On one hand, there is a clear need of increase in the

Joint RES and Distribution Network Expansion Planning under a Demand Response Framework.

renewable generation for environmental reasons. On the other hand, system reliability must be guaranteed while minimizing the costs of supplying the electricity demand. The increasing penetration of stochastic resources in electrical networks calls for a review in uncertainty modeling. In medium- and long-term power system models, it is generally accepted to approximate the demand curve by load levels. The advantage is that this approach lies on the fact that the computational burden is much lower, making the problem tractable. However, the main disadvantage of the considered representation is that, by doing so, information of the sequentiality of individual hours is lost. Furthermore, in power systems with high renewable penetration, the considered modeling procedure to incorporate uncertainty in wind and solar irradiation may not be sufficiently accurate, creating significant distortions in the model results. A common modeling approach to integrate renewables with the screening curve methodology applied in the literature is to use the "net load duration curve." The net demand curve results from subtracting hourly nondispatchable generation output from hourly load. Nondispatchable generation is thus modeled as a deterministic load-modifier or negative load. The use of this procedure does not eliminate the problem, since hours with high demand and high renewable generation and hours with low demand and low renewable penetration would fit the same net demand block. The influence of uncertain factors in the expansion planning is considered by applying scenario-based programming techniques. A multiperiod model for composite power system expansion planning is formulated as a mixed integer programming model.

As the previous discussion reflects, one key issue related to investment decisions for electric power systems with a high penetration of stochastic generation resources and price-dependent resources is the modeling of variability and uncertainty that influences the results of the investment problem. The integration of price-dependent resources, such as DR, calls for a review in the proposed methodology increasing the number of load blocks. A more detailed representation of the demand, wind, and irradiation curves is required to adequately allocate the effects of DR. A total of 144 blocks have been considered in the model formulation. The proposed methodology represents uncertain input data using a set of scenarios, accurately describing the associated uncertainty. The proposed cooptimized expansion planning model is formulated as an instance of stochastic programming where the correlated intraannual uncertainty of demand, wind power, and PV power is characterized by a set of scenarios. A multistage planning framework has been adopted in the formulation of the model, where the annual load curve is discretized into several blocks and increased across the time horizon.

Despite interannual uncertainty may play a critical role in the proposed expansion planning, for the sake of simplicity the considered historical data have been applied for all years included in the planning horizon. The model could be enhanced by introducing interannual wind speed and irradiation patterns, altering the considered scenarios for each stage of the expansion planning model.

3.2 METHOD BASED ON LOAD, WIND, AND IRRADIATION CURVES

In this chapter, the criteria used to define the considered load blocks, which will allow the addressing of the arising challenges of the integration of DR in electric power systems with high renewable penetration, is firstly described. Then, the methodology adopted to model the uncertainty associated to demand, wind speed, and solar irradiation will be described in detail.

In order to adequately accommodate the increasing amount of renewable generation in the expansion planning, a novel approach to model load levels in electric power systems with high wind and PV penetration has been considered, dividing the traditional load duration curve in a tractable amount of blocks. The considered criteria to split the load, wind, and solar irradiation curves are: quarter, working/nonworking day, and day/night. While quarter and working/nonworking day criteria are fixed criteria, day/night criterion has been considered based on solar irradiation. Therefore, an hour where solar irradiation is higher than zero is considered as day in the model, while an hour without solar irradiation is assigned to the corresponding night period. Doing so, the best possible integration of PV technologies within the burdens of the proposed methodology is guaranteed. Historical data of wind and PV power capacity factors (or wind speed and solar irradiation) are used throughout the same period considered for the demand. Wind and PV capacity factors are considered for all hours included in each demand block. Using the obtained data, wind and PV duration curves for each block are built, arranging the data from higher to lower values in order to be jointly represented with the demand (Fig. 3.1). In order to reduce the loss of information on the sequence of individual hours when composing the load duration curve, load, wind, and irradiation curves have been split into 16 different load curves, based on quarters (q), weekend/working day (r), and day/night criteria (n), containing each curve three blocks (b). We use historical data of wind speed and irradiation throughout the same period considered for the demand. Then, for all the hours allocated to each demand block, we consider the corresponding wind and photovoltaic power capacity factors. The uncertainties in the demand, wind speed, and solar irradiation within each demand block are represented by considering different demand factor levels. To do so, we build the cumulative distribution function (cdf) of the demand, wind, and PV factors within each demand block. Then, this cdf is divided into a selected number of segments (three in the proposed paper), each one with an associated probability. In order to simplify the formulation of the problem, each stage is considered to be defined by 3 years.

The modeling of uncertainty in a distribution system for each single load block is based on the load, wind, and solar irradiation curves. Hourly historical demand data is arranged from higher to lower values keeping the correlation between the different hourly data of wind and PV productions. The load duration curve is approximated using demand blocks. In order to accurately include peak demand in the

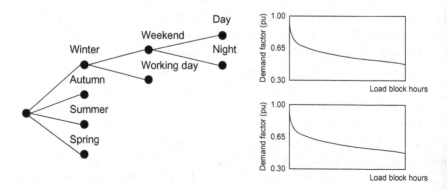

Figure 3.1 Load, wind, and irradiation curves.

model, the length of the demand blocks varies along the load duration curve. For each demand block, wind and PV productions are arranged from higher to lower values in order to be jointly represented with demand. In order to reduce the loss of information on the sequence of individual hours when composing the load duration curve, load, wind, and irradiation curves have been split into 48 different blocks based on quarter (4), weekend/working day (2), and day/night (2) criteria. For each demand block, all combinations of demand, wind power factor, and PV power factor levels are considered. Each combination is assigned a probability within the demand block equal to the probability of the demand factor level times the probability of the wind power factor and the PV power factor levels. For this chapter, 48 demand blocks, 3 factor levels for demand, 3 factor levels for wind power, and 3 levels for PV power have been considered. It is worth mentioning that the methodology proposed in this chapter is related to a central planning context, where distribution assets are both owned and operated by Distribution Companies (DISCOs). Therefore, energy storage system (ESS) is managed to increase the economic benefit of the generation and distribution company. On the contrary, DR reflects the behavior of consumers, aimed at reducing the total payment over the considered time horizon. The impact of both technologies will be analyzed in the proposed case study, outlining the above-mentioned effect (Fig. 3.2).

The modeling of uncertainty in a distribution system for each single load block is based on the load, wind, and solar irradiation curves [40, 41]. The methodology can be described in seven steps.

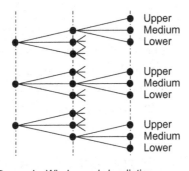

Demand   Wind speed   Irradiation

Figure 3.2 Considered scenario tree for demand, wind speed, and irradiation.

3.2.1 Step 1

Knowing the historical hourly data of demand, wind speed, and solar irradiation, wind speed and solar irradiation can be transformed into power output values using the production models. Then, these data can be expressed as factors (dividing the wind and PV nominal power by the peak demand). In the case of no wind power production historical data in the considered location, historical wind power capacity factors are obtained considering historical data of wind speeds, and transforming these wind-speed data into wind power capacity factors through appropriate wind-speed/wind-power production curves. The methodology to convert wind-speed data into wind power capacity factors is described in chapter "Renewable Power Generation Models."

3.2.2 Step 2

The hourly historical demand data are arranged from higher to lower values keeping the correlation between the different hourly data of wind and PV productions, as depicted in Fig. 3.3.

3.2.3 Step 3

Demand blocks are defined. The load duration curve is approximated using demand blocks. In order to accurately include peak demand in the model, the length of the demand blocks varies along the load duration curve. The first demand block is narrower than the second and the third in order to capture the effect of the peak demand, which usually has a great impact on system-wide decisions and must be adequately represented (Fig. 3.4).

3.2.4 Step 4

The cdf of the demand factors in each demand block is used to represent uncertainty. The cdf is divided into a known number of segments, each of them with a determined probability. Then, the demand factor levels are built with the average values of the demand factors, as depicted in Fig. 3.5.

3.2.5 Step 5

Next, wind and PV power production are modeled. PV power production is partially correlated with the demand. Wind is usually anticorrelated with the demand of the system. All of them, demand,

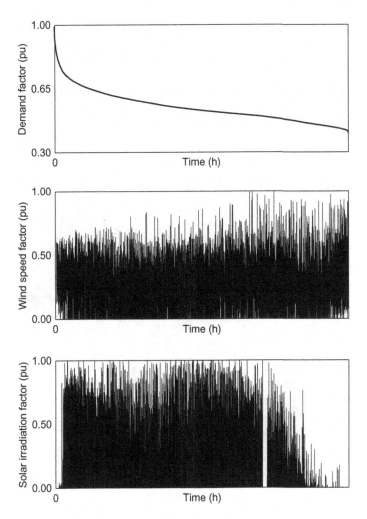

Figure 3.3 Demand curve.

wind power production, and PV power production have to be jointly represented to capture the combination of both effects. Alternative approaches such as the net demand approach (demand minus non-dispatchable generation) do not capture the relevance of this effect, which increases with the renewable penetration. Historical data of wind and PV power capacity factors (or wind speed and irradiation) are used throughout the same period considered for the demand. For all hours included in each demand block, wind and PV capacity factors are considered. Using the obtained data, wind and PV duration curves

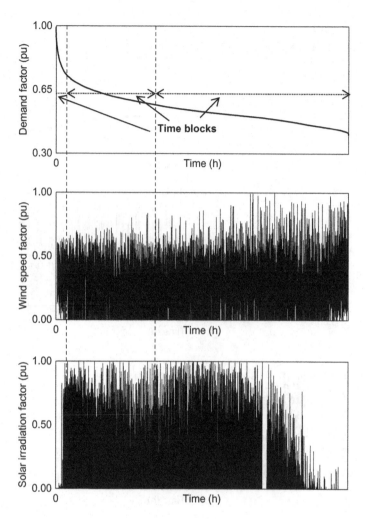

Figure 3.4 Demand curve with demand blocks.

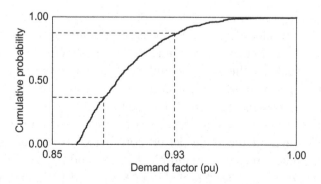

Figure 3.5 Cumulative distribution function of demand factors.

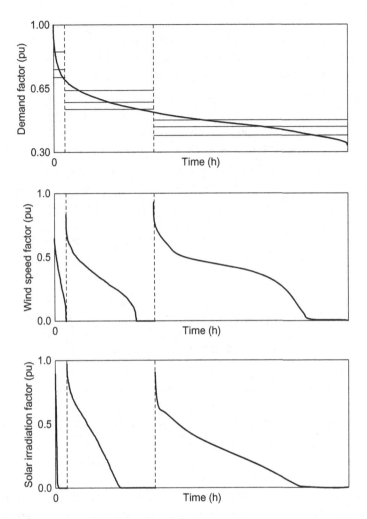

Figure 3.6 Wind and PV power for each demand block.

for each block are built, arranging the data from higher to lower values in order to be jointly represented with the demand. This procedure is represented in Fig. 3.6.

3.2.6 Step 6

Wind and PV duration curves are approximated by a set of wind and PV power capacity factor levels. The cdf of the wind and PV capacity factors in each demand block is used to represent uncertainty. The cdf is divided into a known number of segments, each of them with a determined probability. The procedure to determine the factor levels is

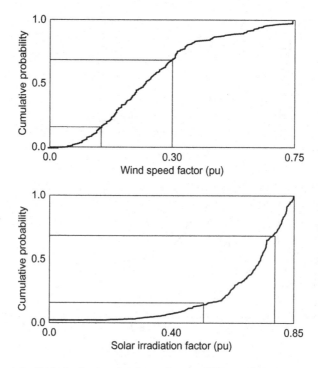

Figure 3.7 Cumulative distribution function of wind power factors and PV power factors.

the same as the one explained in Step 4 but, in this case, for wind and PV power (Fig. 3.7).

3.2.7 Step 7

For each demand periods, all combinations of demand, wind power factor, and PV power factor levels are considered. Each combination is assigned a probability within the demand block equal to the probability of the demand factor level times the probability of the wind power factor and the PV power factor levels. For example, if there are 3 demand blocks, 3 factor levels for demand, 3 factor levels for wind power, and 3 levels for PV power, a total of 81 operating conditions $(3 \times 3 \times 3 \times 3)$ are obtained. Final representation of demand, wind power capacity, and PV power capacity for a single load level is presented in Fig. 3.8. The number of load levels considered to represent the yearly demand, the number of demand blocks considered to adjust the load duration curve within each load level, as well as the number of demand factor levels, wind power capacity factor levels, and PV power capacity factor levels used to represent the uncertainties

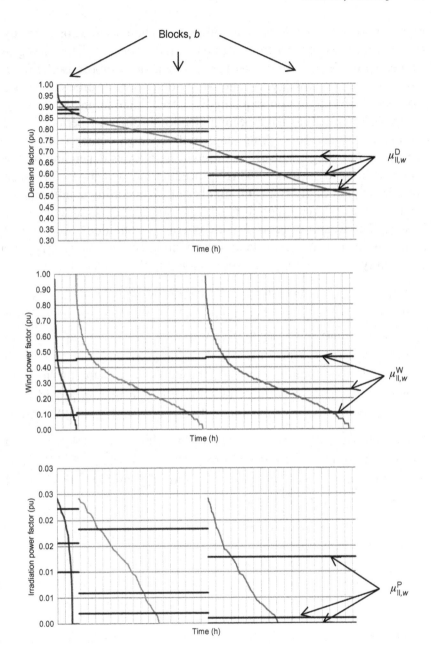

Figure 3.8 Load blocks.

in demand, wind and PV power production should be selected taking into account the nature of the study to be carried out. Selecting a large number of load levels, blocks, and levels may result in an intractable problem, while a too small number may lead to a poor representation of the considered conditions. The relevance of correctly incorporate time-dependent resources and stochastic technologies in the expansion planning calls for an adequate compromise between tractability and simplicity. As previously described, in order to reduce the loss of information on sequentially of individual hours, load, wind, and irradiation curves have been split into 48 different periods based on quarter, weekend/working day, and day/night criteria. Fig. 3.8 represents the assignment to the different load levels depending on the above-mentioned criteria. This procedure is repeated for each considered period, given a total number of 27 operating conditions for each of the 240 considered blocks (5 stages $\times$ 16 load curves $\times$ 3 blocks). The weight of each of these operating conditions is computed as the number of hours in the corresponding demand block times the probability of each operating condition within the demand block.

For each load level ll, each scenario w comprises an average demand factor $\mu_{ll,w}^{D}$, a maximum level of wind power generation $\mu_{ll,w}^{W}$, and a maximum level of PV power generation $\mu_{ll,w}^{\theta}$. Nodal demands in each scenario are equal to the product of the forecasted values and the demand factor $\mu_{ll,w}^{D}$ throughout the planning horizon. Moreover, for each scenario w, based on the information provided by $\mu_{ll,w}^{\theta}$ are converted to maximum levels of wind and PV power generation, $\hat{G}_{i,t,ll,w}^{W}$ and $\hat{G}_{i,t,ll,w}^{\theta}$, respectively. This assumption is deemed adequate considered the size and orography of the island. For the sake of simplicity, we consider that $\mu_{ll,w}^{W}$ and $\mu_{ll,w}^{\theta}$ are identical for all candidate nodes and all considered time stages.

CHAPTER 4

Demand Response Modeling

4.1 DEMAND RESPONSE MODELING

Electrical systems have been historically designed to ensure that demand is met in the most cost-effective way, subject to any possible constraint. Under this design, an increasing amount of network capacities and installed generation would only be dispatched and used during peak periods, which represent few hundreds of hours per year. Electric power systems have three important characteristics. First, because electricity cannot be stored economically, the supply of and demand for electricity must be maintained in balance in real time. Second, grid conditions can change significantly from day-to-day, hour-to-hour, and even within moments. Demand levels also can change quite rapidly and unexpectedly, and resulting mismatches in supply and demand can threaten the integrity of the grid over very large areas within seconds. Third, the electric system is highly capital-intensive, and generation and transmission system investments have long lead times and multi-decade economic lifetimes. Demand side management (DSM), demand response (DR), and energy efficiency allow consumers to actively participate in energy management system, reducing the need for additional generation. Of emerging relevance to resource planning is the potential future role of DR to facilitate the integration of variable and intermittent sources of generation, such as wind and solar sources. While this is largely an operational issue, it also has relevance to long-term resource planning where decisions are being made about when and how to integrate renewables to meet renewable portfolio standards requirements. The role of DR becomes in this case crucial to overcome RES fluctuations with the increasing penetration of intermittent renewable energy sources in the electrical grid.

Demand Response is becoming a part of the system operations in the smart grid driven restructured power system around the world. DR programs aim at defining appropriate price signals to promote changes in electric usage by end users from their original consumption pattern in response to electricity price over time. DR can be also be defined as an incentive payment to reduce electricity consumption in times of high-energy prices, increasing the electricity consumption at times of low prices. However, the influence of DR programs has not yet been exhaustively considered in expansion planning modeling. Any technique to reduce and shift demand to other periods in the day is beneficial for asset utilization and capital investment, which in turn can reduce tariffs. A common approach to managing the load curve involves introducing peak and off-peak tariffs, usually corresponding to day and night, to encourage consumers to modify their consumption behaviour so as to lower peak load. This approach can be useful for customers as well as for utilities. In small island communities with a good understanding of how individual customers contribute to overall demand, it is easier to manage this demand than to achieve large interconnected networks. The increasing penetration of intermittent renewable energy sources (eg, solar, wind) and the development of advanced communication technologies and infrastructure give rise to questions on how loads can be made more flexible in order to optimize the use of resources and assets. Smart grids and demand-side participation (especially with the information provided by smart meters) offer new opportunities to redesign and optimize the power system, increasing its overall efficiency.

In this chapter, DSM and DR will be defined and categorized, focusing on DR activities. Different approaches to transmit price signals to the incumbents and to model DR are also reviewed. Finally, specific studies are commented on.

As usually described in the literature, demand side participation typically encompasses two concepts: DR and DSM. DR, as well as DSM, aims at modifying the demand profile, making consumers aware of the different prices of electricity throughout the day. Details of both approaches can be found in Refs. [42–44]. DSM, or load management, has been used by the power industry over the last decades aiming at reducing or removing peak load, hence deferring the installation of new capacities and distribution facilities [45]. DSM is therefore characterized by a "top-down" approach where the utility decides to implement measures on the demand side to increase the efficiency of the energy system. DSM represents the planning, implementation, and monitoring of those utility activities designed to influence customer use of electricity in ways that will produce desired changes in the utility's load shape, that is, changes in the time pattern and magnitude of a utility's load. Different DSM solution have been put in place by electric utilities, including load management, new uses, strategic conservation, electrification, customer generation, and adjustments in market share. In general terms, the following critical aspects of energy planning are included in DSM:

- The scope of DSM is to influence the customer's use of energy in order to effectively face the system requirements.
- In order to evaluate the impact of the proposed DSM measures, results will be compared against non-DSM alternatives, such as generating units, purchased power, or supply-side storage devices. It is at this stage of evaluation that DSM becomes part of the integrated resource planning process.
- DSM relies on the identification of consumer's behavior, not on the expectation of how consumers should react.
- DSM measures will be economically evaluated based on how they influence costs and benefits throughout the day, week, month, and year through the variation on the load shape.

DR, on the contrary, represents a "bottom-up" approach. DR refers to mechanisms to manage the demand from customers in response to supply conditions, for example, having electricity customers reduce their consumption at critical times or in response to market prices. Being a resource that can be quickly deployed, DR is recently growing in importance, due to the tight supply conditions in certain regions. Two types of DR can be broadly identified: incentive-based DR and time-based DR.

Incentive-based programs are based on customer response to incentives paid by the electricity utility in times of high electricity price in order to motivate customers to reduce their consumption. Incentive-based DR includes direct load control, interruptible/curtailable rates, demand bidding/buyback programs, emergency DR programs, capacity market programs, and ancillary services market programs. On the other hand, time-based DR relies on customer's choice to decrease or change their consumption in response to changes of electricity's price. Time-based DR alternatives include time-of-use (TOU) rates, critical peak pricing (CPP), and real-time pricing (RTP). The formulation proposed in this chapter relies on the last, incorporating in the problem using the reaction of the consumers to energy prices at each considered block, while dismissing other possible solutions such as TOU, CPP, or Interruptible Demand. DR programs may help in reducing the total cost of maintaining system balance by modifying consumption, particularly when the opportunity cost of energy is high.

Among the objectives of DR programs (depending on its implementation), the following can be highlighted:

* Peak shifting or clipping: Reduce peak demand at specific periods, reducing therefore the need for installing additional generation units.
* Valley filling: Increasing the load during off-peak hours and therefore.
* Load shifting: Shifting peak period loads to off-peak hours.
* Electricity savings: Decreased overall electricity consumption throughout the year.

DR offers mechanisms to electricity users to modify the electric use by end-use customers in response to changes in the price of electricity over time, or to give incentive payments designed to induce lower electricity use at times of high market prices or when grid reliability is jeopardized. The DSM programs can be broadly divided into incentive-based and price (time)-based programs

4.1.1 Incentive-Based Programs (IBP)

Incentive-based DR programs are usually based on economic incentives and are often not included in the electricity rates. A further division can be established among IBP, namely, market based and classical programs. Market-based IBP includes Emergency DR programs, Demand Bidding, Capacity Market, and Ancillary services market. Classical IBP include Direct Load Control programs and Interruptible/Curtailable programs. While in classical programs, consumers receive payments for their participation, in market-based programs participants are rewarded money for their performance depending on the amount of load reduction during critical conditions.

While in Direct Load Control programs, utilities have the ability to remotely shut down participant equipment; customers participating in Interruptible/Curtailable Programs will receive incentive upfront payments when asked to reduce their load to predefined values. Demand Bidding (also called Buyback) allows consumers to bid load reductions in electricity wholesale market. On the contrary, in Emergency DR programs, consumers are paid incentives for their measured load reductions during emergency conditions. Capacity market programs provide capacity payments to customers for their agreement to curtail when directed. Finally, Ancillary services market programs allow consumers to offer load reductions in the intraday energy markets.

4.1.2 Price-Based Programs (PBP)

In PBP programs, dynamic pricing rates are offered to the consumers by changing electricity sale price in different time periods. In this context, the price of electricity in peak periods is significantly higher than off-peak periods, flattering the demand curve. RTP, TOU, Extreme Day Pricing (EDP), CPP, and Extreme Day CPP (ED-CPP) are the programs in this category. In TOU, unit price of electricity changes during certain time periods. Calculation of the unit price is done based on average electricity consumption in different periods. EDP, CPP, and ED-CPP extra pricing are done only for special conditions. In RTP, customers are informed about the prices on a day-ahead or hour-ahead basis. There is a general consensus among economists suggesting that RTP programs are the most direct and efficient DR programs suitable for competitive electricity markets and should be the focus of policymakers. A mature advance-metering infrastructure makes possible the transmission of adequate price signals to the consumers, revealing their implicit elasticity. Authors in Refs. [46] and [47] analyze different signals for DR: first, dynamic pricing and time-varying prices, second, interruptible and voluntary load shedding, and, third, the inclusion of the demand in ancillary service mechanisms. Dynamic pricing refers to retail prices varying with real-time system conditions and requires hourly meters to be implemented, while TOU pricing refers to prices set in advance but varying over the day to capture expected impacts of changing electricity conditions. Real-time prices include the result of the market clearing process and depend

---

Demand response programs

- Incentive-based programs (IBP)
  - Classical
    - Direct control
    - Interruptible / curtailable programs
  - Market based
    - Demand bidding
    - Emergency DR
    - Capacity market
    - Ancillary services market
- Price-based programs (PBP)
  - Time of use (ToU)
  - Critical peak pricing (CPP)
  - Extreme day CPP (ED-CPP)
  - Extreme day pricing (EDP)
  - Real time pricing (RTP)

Figure 4.1 Classification of demand response programs.

(theoretically) on the marginal cost of the system, giving customers an incentive to reduce demand during high wholesale price periods. Real-Time Pricing reflects the current conditions and provides the best available signal about the marginal value of power at a location. Under a TOU pricing methodology, customers have no incentive to reduce their energy consumption during high wholesale price periods. The formulation proposed in this chapter foresees the implementation of RTP, completely incorporating consumers in the outcomes of the hourly dispatch and taking advantage of the elasticity of the demand side (Fig. 4.1).

4.2 FORMULATION

The inclusion of DR in the problem formulation requires an adequate treatment of uncertainties in the system and load levels. In medium- and long-term power system modeling, demand has been traditionally approximated by load levels, in order to make the problem tractable. In an attempt to accommodate the increasing amount of RES in existing power systems and to provide an adequate framework for DR impact in existing power system models, load period approach and clustering approach have been recently applied to accurately represent load curve. Further details on the methodology adopted to model the uncertainty associated to demand, wind speed, and solar irradiation in a distribution system can be found in the corresponding chapter.

DR has been introduced in the model considering elastic demand functions calibrated by load levels. The demanded energy in every load level (4.1) has been expressed as a function of elasticity, demand, and prices for the incumbent load levels included in the load-shifting horizon and average price (4.2). The slope of the demand function is determined by the elasticity considerations already discussed. At least some of the customers are considered to participate in a real-time pricing methodology, where consumers face risks associated to spot pricing. The considered elasticities represent the participation of consumers in such tariff modeling, where smaller elasticities correspond to lower participation level.

$$d_{i,t,\text{ll},w} = D_{i,t,\text{ll},w} + \sum_{\text{lb}} \xi_{\text{lb,ll}} \frac{(D_{i,t,\text{ll},w}\Delta_{\text{ll}} + D_{i,t,\text{lb},w}\Delta_{\text{lb}})}{\Delta_{\text{lb}} + \Delta_{\text{ll}}} \frac{(c^{\text{SS}}_{t,\text{lb},w} - \overline{C^{\text{SS}}_{t,\text{ll},w}})}{\overline{C^{\text{SS}}_{t,\text{ll},w}}} \quad (4.1)$$

$$\overline{C^{\text{SS}}_{t,\text{ll},w}} = \frac{\sum_{\text{lb}} C^{\text{SS}}_{t,\text{lb},w}\Delta_{\text{lb}}}{\sum_{\text{lb}}\Delta_{\text{lb}}} \quad (4.2)$$

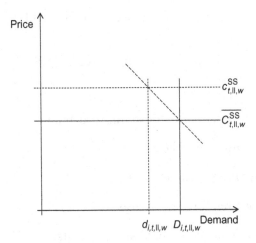

Figure 4.2 Elastic demand function.

The reference price–quantity pair composed of the weighted average price and the fixed demand level $\{p_{i,\text{ll}}, D_{i,\text{ll}}\}$ is considered to be the base point of the linear demand function for each load level. The price elasticity assumptions determine the slope of the demand function with own-price elasticities $(\xi_{\text{ll,ll}})$ and cross-price elasticities $(\xi_{\text{lb,ll}})$ being exogenously provided, based upon values from the literature. Own-price elasticity refers to how energy consumers react to every single load level, considering the average price of the incumbent load levels. Cross-price elasticity (also known as elasticity of substitution) refers to the consumer's reaction to the prices in other load levels. The addition of price elasticities results in the DR function $d_{i,\text{ll}}$ (4.1), which expresses the quantity demanded $(d_{i,\text{ll}})$ as function of the relative deviations of load level prices from the reference level. Fig. 4.2 shows the portion of that function that relates price to demand in its own period. In Ref. [48] a clear analysis of DR sensitivities to participation rates and elasticities is presented.

4.3 METHODOLOGIES TO INCLUDE SHORT-TERM DR INTO EXPANSION PLANNING

With the traditional approach, in general, the distribution network must be jointly optimized together with generation investments. Here the objective is to maximize consumer's welfare (utility function minus costs):

$$\max\{U(D) - \text{FGC} - \text{VGC} - \text{NC}\} \qquad (4.3)$$

where:

$U(D)$ = utility function of consuming a demand D;

FGC = generation fixed costs;

VGC = generation variable costs;

NC = network reinforcement costs;

which can be basically considered as fixed costs.

The demand is assumed to be given and generation planning is also prescribed from the outset, network planning becomes the typical minimization of generation operation costs via network reinforcement:

$$\min\{\text{VGC} + \text{NC}\} \qquad (4.4)$$

Within a competitive approach the entity in charge of network planning must apply the following optimization criterion in order to identify the network reinforcements that must be proposed to the regulatory entities for authorization:

$$\max\{\text{Net benefit of consumers} + \text{Net benefit of generators}\} \qquad (4.5)$$

However, when DR is integrated in the expansion planning model, the minimization of costs partially disregard the benefits consumers receive from the modification of their electricity consumption [11]. Different methodologies have been defined in literature [49−55] to adequately account for DR in both network and generation expansion planning modeling. The total surplus maximization problem (4.6) can be seen as:

Max net social welfare =

$$\max\left\{ \begin{array}{l} \sum_{t,\text{ll},w}\left((1+I)^{-(t-1)} + \dfrac{(1+I)^{-(n_T-1)}}{I}\right)\left[\pi_w p_{t,\text{ll},w}\right] - \sum_t \dfrac{(1+I)^{-(t-1)}}{I}c_t^{\text{I}} \\ -\sum_t\left[(1+I)^{-(t-1)}(c_t^{\text{M}} + c_t^{\text{E}} + c_t^{\text{ST}} + c_t^{\text{U}})\right] - \dfrac{(1+I)^{-(n_T-1)}}{I}(c_{n_T}^{\text{M}} + c_{n_T}^{\text{E}} + c_{n_T}^{\text{ST}} + c_{n_T}^{\text{U}}) \end{array} \right\}$$

$$(4.6)$$

CHAPTER 5

Energy Storage Systems Modeling

As a result of the integration of renewable resources and distributed systems in insular grids, in particular, the electricity generated using renewable sources (such as solar and wind generation), which do not operate all the time and have huge fluctuations due to their stochastic nature, is difficult to adjust in response to the demand needs. Therefore, storage systems are needed to avoid stability problems. The main advantage of ESS systems is the release of additional capacity to the grid when it is valuable, their numerous applications will strengthen power networks and maintain load levels even during critical service hours. As a result, ESS systems represent the critical link between the energy supply and demand chains, standing as a key element for the increasing grid integration of renewable energies. Furthermore, ESS will enable the smart grid concept to become a reality. Special emphasis is given to ESS on islands, studying their particular requirements, the most appropriate technologies and the generic storage mathematical model in terms of distribution network expansion planning.

Some of the critical concerns to consider for the deployment of ESS are: the optimal sizing of ESS, the project financial sustainability, system complexity and integration, end-user buy-in (financially and politically), and systematic deployment strategy. It is difficult to select only one single best storage solution as experience showed that storage is not always necessarily appropriate for island electricity systems. By its nature, storage can be very valuable in transmission systems with capacity constraints and in distribution systems with low power quality at the end of its network. However, it may not be adequate for solving chronic supply shortage or poorly performing transmission and distribution systems [56]. In consequence, the main purpose of ESS utilization in island grids is not to act as a back-up of basic diesel generation, since it is a well-known fact that storage devices definitely enhance diesel efficiency. The present main focus of EES integration goes forward for the sake of the increased commitment for renewable

Joint RES and Distribution Network Expansion Planning under a Demand Response Framework.

integration and grid code fulfilment in isolated grids in medium- and long-term distribution and generation expansion planning procedures. Additionally, islands have special features that distinguish them from mainland interconnected grids.

A range of storage solutions and integration case studies for island applications has been reviewed with different storage technologies, where most of them are battery technologies from low and high power capacity being relevant the most matured ones in the order of importance (conventional and advanced lead-acid battery, flow batteries, Lithium-Ion, and NAS). In addition, there exits pumped storage plants which requires long construction times and high capital expenditure. The modalities of this storage system are: upper-lower reservoirs, high dam hydro plants with storage capability, underground pumped storage, or open sea reservoirs. In spite of a major technological advancement on ESS technologies and their application in highly reinforced grid and distributed generation integrated grids, there are still many challenges related to island applications, especially when the isolated system operation is combined with high renewable energy participation scenarios. No similarity or uniformity on ESS application could be verified with total certainty since it is dependent on the scale and type of application requirements. However, it can be outlined that there are appreciable trends in the majority of storage applications in a broad sense in terms of renewable integration, power control, time shifting, smoothing or quality of supply and basic supply services.

Finally, practical benefits are described in terms of technical and economic aspects and technologies. Nevertheless, the specifics about remaining challenges in each case should be closely considered. As a conclusion it can portrayed that any solution provided to meet grid requirements should be aligned to the specific island needs and grid requirement, complexity of integrated items, abundance of the resources, the scale of economy and the type of storage technology to be deployed.

5.1 BENEFITS DERIVED FROM THE USE OF ESS

Potential applications of ESS include a wide range of aspects, covering the full spectrum of energy power system requirements ranging from larger scale, generation, and transmission-related systems to those primarily related to the distribution network. ESS benefits can be classified into two generic categories: technical and economic benefits. On one hand, high-power ESS provide reliability, safety, and productivity, that is, it provides technical benefits. On the other hand, high-energy ESS would help improving profitability, that is, it secures economic benefits to the power system stakeholders.

5.1.1 Technical Benefits

The most relevant technical benefits derived from the integration of ESS in the electric power systems are mentioned below:

1. ESS could be used to time-shift electric energy generated by renewables, playing an important role in the integration of renewable energy into the grid [57].
2. As a result of the integration of ESS, a more efficient use and contribution of renewable energy is expected, also fomenting the use of distributed energy supply options in grids.
3. Base-load generation plants are, in general, not designed to provide variable output. Storage may provide attractive solution to these drawbacks by setting the optimal operation point, rather than firing standby generators. In addition to that, ESS have superior part-load efficiency [58].
4. Storage output can be changed rapidly giving a ramping support and black-start to the grid [58]. ESS can be used for energy management for large-scale generations such as spinning reserves, ramping services, load following, and load leveling.
5. ESS could improve the performance of the Transmission and Distribution system by providing the utilities the ability to increase energy transfer and stabilize voltage levels, reducing congestion in the system [59].
6. Energy storage can be used as a solution for improving grid service reliability, helping Distribution Companies (DISCOs) to partially "ride-through" a power disruption.
7. Energy storage can benefit utilities or independent system operators can benefit from ESS by deferring transmission and distribution upgrades.
8. ESS can serve as a standby power source for substations on-site and distribution lines, or to add transformers.
9. Other nonelectrical energy uses may benefit from the deployment of ESS, like transportation and heat generation.

5.1.2 Economic Benefits

Some of the economic benefits derived from the integration of ESS in the electric power systems are mentioned below:

1. ESS can successfully contribute to the cost reduction from electricity consumers.
2. In general, off-peak electricity is cheaper compared to high-peak electricity, also benefitting the seller of electricity. ESS can be used to acquire inexpensive electricity available during low demand periods to charge the storage plant, so that the low priced energy can be used or sold at a later time when the price for electricity is high.
3. ESS could be used to offset the need to purchase and install new generation, avoiding unnecessary additional cost burdens for generators.
4. Economic development and employment opportunities for many countries will be possible through the integration on ESS in the system.
5. It will allow more efficient use of renewable time-shifting electric energy generated by these generators, increasing economic benefits derived these technologies.
6. In a deregulated environment, ESS may help utilities to avoid transmission congestion charges, which are very expensive [59].
7. The need for Transmission and Distribution capacity upgrades may be reduced as a consequence of the integration of ESS, thus minimizing unnecessary investments.
8. The use of ESS tends to lower GHG emissions, reducing the environmental impact of energy requirements. However, this cost reduction is specific to the resource and varies greatly between technologies.

5.2 ESS TECHNOLOGIES

A list of the most relevant ESS technologies [59] is as follows:

- Pumped hydro energy storage (PHES).
- Compressed air energy storage (CAES).
- Small-scale compressed air energy storage (SS-CAES).
- Thermal energy storage (TES).
- Hydrogen energy storage system (HESS).

- Chemical energy storage system/batteries (BESS).
  - Advanced and lead–acid batteries.
  - Nickel–cadmium batteries.
  - Nickel–metal hydride batteries.
  - Lithium–ion batteries.
  - Sodium–sulfur batteries.
  - Sodium nickel chloride batteries.
- Flow batteries energy storage (FBES).
  - Vanadium redox flow battery.
  - Zinc–bromine redox flow battery.
  - Polysulfide bromide Regenesys flow battery.
- Flywheel energy storage system (FESS).
- Supercapacitors energy storage (SCES).
- Superconducting magnetic energy storage (SMES).
- Energy storage in substitute natural gas (SNG).
- Electric vehicles (EVs).

5.3 STORAGE UNIT MODEL

In modern power systems, storage devices have been grown rapidly. Energy storage units are integrated to energy distribution systems (EDS) to meet several purposes such as real-time power demand, smoothing output power of renewable energy resources (RES), improving power system reliability, and being economically efficient [60]. Consequently, a generic storage system is modeled in this report and mathematically represented through (5.1)–(5.4):

$$\underline{G}^{ST} u_{i,t,\mathrm{ll},w}^{ST,\,prod} \leq g_{i,t,\mathrm{ll},w}^{ST,\,prod} \leq \overline{G}^{ST} u_{i,t,\mathrm{ll},w}^{ST,\,prod} \tag{5.1}$$

$$\underline{G}^{ST} u_{i,t,\mathrm{ll},w}^{ST,\,store} \leq g_{i,t,\mathrm{ll},w}^{ST,\,store} \leq \overline{G}^{ST} u_{i,t,\mathrm{ll},w}^{ST,\,store} \tag{5.2}$$

$$u_{i,t,\mathrm{ll},w}^{ST,\,prod} + u_{i,t,\mathrm{ll},w}^{ST,\,store} \leq y_{i,t}^{ST} \tag{5.3}$$

The power supplied or stored by storage units is limited by the minimum between their capacities and the maximum capacity (5.1) and (5.2). Finally, in (5.3), binary variables $u_{i,t,\mathrm{ll},w}^{ST,\,prod}$ and $u_{i,t,\mathrm{ll},w}^{ST,\,store}$ are defined to avoid producing and storing energy simultaneously. Energy storage units are integrated to EDS to meet several purposes such as real-time power demand, smoothing output power of RES, improving power system reliability and being economically efficient.

Consequently, a generic storage system is modeled and mathematically represented through (5.4):

$$\sum_{n,f} \left[\Delta_{ll} \left(\eta_i^{ST,\ store} g_{i,t,ll,w}^{ST,\ store} - \left(\frac{1}{\eta_i^{ST,prod}} \right) g_{i,t,ll,w}^{ST,\ prod} \right) \right] = 0 \qquad (5.4)$$

$$\forall i \in \Omega^N; \forall t \in T; \forall ll \in LL; \forall ST \in \Omega^{ST}; \forall w \in W$$

In medium- and long-term power system models, it is a common approach to approximate the demand curve by load levels in order to make the models computationally tractable. However, in such an approach, the chronological information between individual hours is lost (and therefore also the transition's function of the ESS). The proposed approach considering different demand blocks following the above-described criteria allows better integrating chronological information in power system models, thereby resulting in a more accurate representation of system outcomes such as electricity prices and total cost. In order to adequately represent the ESS transition, (5.4) is formulated to be accomplished at an n,f level (day/night, load block). In this book, storage units are possible device candidates in the island of La Graciosa network, as shown in equations and results in chapter "Case Study."

CHAPTER 6

Optimization Problem Formulation

A great variety of models has been proposed in literature for distribution and generation system planning. In this chapter, the mathematical formulation for the optimization problem of distribution network expansion planning is presented. Main references have been detailed in chapter "Renewable Power Generation Models," providing a summary description of the performed activities. The proposed model is built on the distribution system planning models described in Refs. [3, 8, 19, 25] wherein: (1) a multistage planning framework is adopted, (2) a discretization of the annual load curve into several load levels is used to characterize the demand, (3) radial operation of the distribution network is explicitly imposed, (4) an approximate network model is used, (5) the costs of losses are included in the objective function, and (6) several investment alternatives exist for each asset. The presented formulation includes a novel model to accurately include demand response (DR) and energy storage systems (ESS) in the joint distribution network and generation expansion planning. Additionally the impact of ESS on both resource plans and its combination with DR has not yet been investigated. The model stresses the importance of integrating DR to time-varying prices (real-time prices) into those investment models. In this regard, own-price and cross-price elasticities are included in order to incorporate consumers' willingness to adjust the demand profile in response to price changes. Timing, location, and sizing of storage units, DG units and distribution assets are modeled. The model proposes social welfare maximization to adequately incorporate price-dependent resources. Besides, different scenarios which represent different operating conditions for demand, wind, and irradiation are modeled. TA scenario-based stochastic programming framework is proposed to model the correlated uncertainty characterizing demand and renewable-based power generation. The associated deterministic equivalent is formulated as a mixed-integer linear program suitable for commercially available software.

Joint RES and Distribution Network Expansion Planning under a Demand Response Framework.

6.1 OBJECTIVE FUNCTION

According to the information shown in chapter "Energy Storage Systems Modeling," the objective function (4.4) has been modified, maximizing the net social welfare of the system. The objective function represents the total payment of the consumers minus the present value of the total cost, which consists of 6 cost terms related to: (1) investment, (2) maintenance, (3) production, (4) losses, (5) unserved energy, and (6) storage.

Max net social welfare =

$$
\max \left\{ \begin{array}{l} \sum_{t,\mathrm{ll},w} \left((1+I)^{-(t-1)} + \frac{(1+I)^{-(n_T-1)}}{I} \right) \left[\pi_{\mathrm{ll},w} p_{t,\mathrm{ll},w} \right] - \sum_t \frac{(1+I)^{-(t-1)}}{I} c_t^{\mathrm{I}} \\ - \sum_t [(1+I)^{-(t-1)} (c_t^{\mathrm{M}} + c_t^{\mathrm{E}} + c_t^{\mathrm{ST}} + c_t^{\mathrm{U}})] - \frac{(1+I)^{-(n_T-1)}}{I} \left(c_{n_T}^{\mathrm{M}} + c_{n_T}^{\mathrm{E}} + c_{n_T}^{\mathrm{ST}} + c_{n_T}^{\mathrm{U}} \right) \end{array} \right\}
$$

(6.1)

The investment cost is amortized in annual payments during the lifetime of the installed equipment, considering that once the component is operated during a time equal to its lifetime, there is a reinvestment in identical equipment, so infinite annual updated payments are used. The remaining costs related to operation are updated and these costs are kept indefinitely, taking into account an infinite series of annual payments [61].

Mathematically, these costs are defined as:

$$
\begin{aligned}
c_t^{\mathrm{I}} = & \sum_{l \in \{\mathrm{NRF,NAF}\}} \mathrm{RR}^l \sum_{k \in K^l} \sum_{(i,j) \in \Upsilon^l} C_{i,j,k}^{\mathrm{I},l} \ell_{i,j} x_{i,j,k,t}^l \\
& + \mathrm{RR}^{\mathrm{SS}} \sum_{i \in \Omega^{\mathrm{SS}}} C_i^{\mathrm{I,SS}} x_{i,t}^{\mathrm{SS}} + \mathrm{RR}^{\mathrm{NT}} \sum_{k \in K^{\mathrm{NT}}} \sum_{i \in \Omega^{\mathrm{SS}}} C_{i,k}^{\mathrm{I,NT}} x_{i,k,t}^{\mathrm{NT}} \\
& + \sum_{p \in P} \mathrm{RR}^p \sum_{i \in \Omega^p} C_i^{l,p} \mathrm{pf} \overline{G}^p x_{i,t}^p + \mathrm{RR}^{\mathrm{ST}} \sum_{i \in \Omega^p} C_i^{\mathrm{I,ST}} pf \overline{G}^{\mathrm{ST}} x_{i,t}^{\mathrm{ST}}; \quad \forall t \in T
\end{aligned}
$$

(6.2)

$$
\begin{aligned}
c_t^{\mathrm{M}} = & \sum_{l \in L} \sum_{k \in K^l} \sum_{(i,j) \in \Upsilon^l} C_{i,j,k}^{\mathrm{M},l} (y_{i,j,k,t}^l + y_{j,i,k,t}^l) \\
& + \sum_{\mathrm{tr} \in \mathrm{TR}} \sum_{k \in K^{\mathrm{tr}}} \sum_{i \in \Omega^{\mathrm{SS}}} C_{i,k}^{\mathrm{M,tr}} y_{i,k,t}^{\mathrm{tr}} \\
& + \sum_{p \in P} \sum_{k \in K^p} \sum_{i \in \Omega^p} C_{i,k}^{\mathrm{M},p} y_{i,k,t}^p + \sum_{p \in P} \sum_{i \in \Omega^{\mathrm{st}}} C_i^{\mathrm{M,st}} y_{i,t}^{\mathrm{st}}; \quad \forall t \in T
\end{aligned}
$$

(6.3)

$$c_t^E = \sum_{w \in \Pi} \pi_{ll,w} pf \sum_{ll \in LL} \Delta_{ll} \left(\sum_{tr \in TR} \sum_{k \in K^{tr}} \sum_{i \in \Omega^{SS}} c_{t,ll,w}^{SS} g_{i,k,t,ll,w}^{tr} \right.$$

$$\left. + \sum_{p \in P} \sum_{i \in \Omega^p} pf C^{E,p} g_{i,t,ll,w}^p \right); \quad \forall t \in T \tag{6.4}$$

$$c_t^R = \sum_{w \in \Pi} \pi_w pf \sum_{ll \in LL} \Delta_{ll} C_{ll}^{SS} \left(\sum_{ll \in LL} \sum_{k \in K^l} \sum_{(i,j) \in \Upsilon^l} R_{i,j,k}^l \ell_{i,j} (f_{i,j,k,t,ll,w}^l + f_{j,i,k,t,ll,w}^l)^2 \right.$$

$$\left. + \sum_{tr \in TR} \sum_{k \in K^{tr}} \sum_{i \in \Omega^{SS}} R_{i,k}^{tr} (g_{i,k,t,ll,w}^{tr})^2 \right); \quad \forall t \in T \tag{6.5}$$

$$c_t^U = \sum_{w \in \Pi} \pi_{ll,w} pf \sum_{ll \in LL} \sum_{i \in \Omega_t^{LN}} \Delta_{ll} C^U d_{i,t,ll,w}^U; \quad \forall t \in T \tag{6.6}$$

$$c_t^{ST} = \sum_{w \in \Pi} \pi_{ll,w} pf \sum_{ll \in LL} \Delta_{ll} \left(\sum_{i \in \Omega^{ST}} C^{ST,prod} g_{i,t,ll,w}^{ST,prod} + C^{ST,store} g_{i,t,ll,w}^{ST,store} \right); \quad \forall t \in T \tag{6.7}$$

where:

$$RR^l = \frac{I(1+I)^{\nu^l}}{(1+I)^{\nu^l} - 1}; \quad \forall l \in \{NRF, NAF\}$$

$$RR^{NT} = \frac{I(1+I)^{\nu^{NT}}}{(1+I)^{\nu^{NT}} - 1}$$

$$RR^p = \frac{I(1+I)^{\eta^p}}{(1+I)^{\eta^p} - 1}; \quad \forall p \in P$$

$$RR^{SS} = \frac{I(1+I)^{\nu^{SS}}}{(1+I)^{\nu^{SS}} - 1}$$

$$RR^{st} = \frac{I(1+I)^{\nu^{ST}}}{(1+I)^{\nu^{ST}} - 1}$$

In (6.2), the investment cost at each stage is formulated as the sum of terms related to: (1) replacement and addition of feeders, (2) reinforcement of existing substations and construction of new substations, (3) installation of new transformers, (4) installation of renewable generators, and (5) installation of storage generic units. Expressions (6.3) model the maintenance costs of feeders, transformers, generators, and storage generic units. Production costs associated with substations and generators are characterized in (6.4). Similar to Ref. [3], the costs of energy losses in feeders and transformers are modeled in (6.5) as quadratic terms. Such nonlinearities can be accurately approximated by a set of tangent lines. This approximation yields piecewise linear functions, which, for practical purposes, are indistinguishable from the nonlinear models if enough segments are used. Expression (6.6) corresponds to the penalty cost of nonserved energy. Finally, (6.7) corresponds to the cost of the energy storage devices.

It is worth emphasizing that, for each time stage, a single binary variable per conductor in the feeder connecting nodes i and j is used to model the associated investment decision making, namely, $x^l_{i,j,k,t}$. In contrast, two binary variables, $y^l_{i,j,k,t}$ and $y^l_{j,i,k,t}$, as well as two continuous variables, $f^l_{i,j,k,t,b}$ and $f^l_{j,i,k,t,b}$, are associated with each feeder in order to model its utilization and current flow, respectively. Note that $f^l_{i,j,k,t,b}$ is greater than 0 and equal to the current flow through the feeder between nodes i and j measured at node i only when the current flows from i to j, otherwise it is 0.

6.2 CONSTRAINTS

At this point, constraints associated with the optimization problem of the joint generation and distribution network expansion planning are formulated.

6.2.1 Integrality Constraints
Investment decisions in new assets are modeled by the following binary variables:

$$x^l_{i,j,k,t} \in \{0,1\}; \quad \forall l \in \{\text{NRF}, \text{NAF}\}, \ \forall (i,j) \in \Upsilon^l, \ \forall k \in K^l, \ \forall t \in T \quad (6.8)$$

$$x_{i,t}^{SS} \in \{0,1\}; \quad \forall i \in \Omega^{SS}, \forall t \in T \tag{6.9}$$

$$x_{i,k,t}^{NT} \in \{0,1\}; \quad \forall i \in \Omega^{SS}, \forall k \in K^{NT}, \forall t \in T \tag{6.10}$$

$$x_{i,k,t}^{p} \in \{0,1\}; \quad \forall p \in P, \forall i \in \Omega^{p}, \forall k \in K^{p}, \forall t \in T \tag{6.11}$$

$$x_{i,k,t}^{st} \in \{0,1\}; \quad \forall i \in \Omega^{st}, \forall k \in K^{p}, \forall t \in T \tag{6.12}$$

For instance, if variable $x_{i,j,k,t}^{l}$ is equal to 1 and l is equal to NRF, the distribution company decides to invest in the replacement of the existing replaceable feeder in branch i-j with the candidate conductor for replacement k at stage t.

Utilization decisions are also modeled by binary variables:

$$y_{j,i,k,t}^{l} \in \{0,1\}; \quad \forall l \in L, \ \forall i \in \Omega_{j}^{l}, \ \forall j \in \Omega^{N}, \ \forall k \in K^{l}, \ \forall t \in T \tag{6.13}$$

$$y_{i,k,t}^{tr} \in \{0,1\}; \quad \forall i \in \Omega^{SS}, \ \forall k \in K^{NT}, \ \forall t \in T \tag{6.14}$$

$$y_{i,k,t}^{p} \in \{0,1\}; \quad \forall p \in P, \ \forall i \in \Omega^{p}, \ \forall k \in K^{p}, \ \forall t \in T \tag{6.15}$$

$$y_{i,k,t}^{st} \in \{0,1\}; \quad \forall i \in \Omega^{st}, \ \forall k \in K^{p}, \ \forall t \in T \tag{6.16}$$

For instance, if variable $y_{i,j,k,t}^{l}$ is equal to 1 and l is equal to NAF, the distribution company decides to use the newly added feeder in branch i-j with candidate conductor k at stage t.

6.2.2 Balance Equations
Constraints (6.17) represent the nodal current balance equations, that is, Kirchhoff's current law.

$$\sum_{l \in L} \sum_{k \in K^{l}} \sum_{j \in \Omega_{i}^{l}} (f_{i,j,k,t,ll,w}^{l} - f_{j,i,k,t,ll,w}^{l})$$

$$= \sum_{tr \in TR} \sum_{k \in K^{tr}} g_{i,k,t,ll,w}^{tr} + \sum_{p \in P} g_{i,t,ll,w}^{p} + \sum_{i \in \Omega^{st}} (s_{i,t,ll,w}^{ST,prod} - s_{i,t,ll,w}^{ST,store}) \tag{6.17}$$

$$+ d_{i,t,ll,w} + d_{i,t,ll,w}^{U}; \quad \forall i \in \Omega^{N}, \ \forall t \in T, \ \forall ll \in LL, \ \forall w \in \Pi$$

These constraints model that the algebraic sum of all outgoing and incoming currents at node i must be equal to 0 for each stage t, load level ll and scenario ω.

6.2.3 Kirchhoff's Voltage Law

The enforcement of the Kirchhoff's voltage law for all feeders in use leads to the following expressions:

$$y^l_{i,j,k,t}[Z^l_{i,j,k}\ell_{i,j}f^l_{i,j,k,t,\text{ll},w} - (v_{i,t,\text{ll},w} - v_{j,t,\text{ll},w})] = 0;$$

$$\forall l \in L, \; \forall i \in \Omega^l_j, \; \forall j \in \Omega^N, \; \forall k \in K^l, \; \forall t \in T, \; \forall l \in \text{LL}, \; \forall w \in \Pi \tag{6.18}$$

These constraints are imposed for all types of lines, such as an existing fixed feeder, existing replaceable feeders, new replacement feeders, and newly added feeders. Note that constraints (6.18) are nonlinear.

6.2.4 Voltage Limits

The nodal voltage modules are limited by upper and lower limits. Mathematically, these constraints are formulated as follows:

$$\underline{V} \leq v_{i,t,\text{ll},w} \leq \overline{V};$$

$$\forall i \in \Omega^N, \; \forall t \in T, \; \forall l \in \text{LL}, \; \forall w \in \Pi \tag{6.19}$$

6.2.5 Capacity Limits for Feeders

The current flow is restricted by the maximum capacity of the feeders. This is formulated as follows:

$$0 \leq f^l_{i,j,k,t,\text{ll},w} \leq y^l_{i,j,k}\overline{F}^l_{i,j,k};$$

$$\forall l \in L, \; \forall i \in \Omega^l_j, \; \forall j \in \Omega^N, \; \forall k \in K^l, \; \forall t \in T, \; \forall ll \in \text{LL}, \; \forall w \in \Pi \tag{6.20}$$

Constraints (6.20) establish the maximum current flow that can be transported through the feeders in use. If a feeder is not used, then the current flow is 0.

6.2.6 Capacity Limits for Transformers

The current supply by substations depends on the number of transformers, which have a maximum current value that can be supplied.

$$0 \leq g^{\text{tr}}_{i,k,t,\text{ll},w} \leq y^{\text{tr}}_{i,k,t}\overline{G}^{\text{tr}}_{i,k}$$

$$\forall \text{tr} \in \text{TR}, \; \forall i \in \Omega^{\text{SS}}, \; \forall k \in K^{\text{tr}}, \; \forall t \in T, \; \forall ll \in \text{LL}, \; \forall w \in \Pi \tag{6.21}$$

Constraints (6.21) set the upper bounds for current that can be supplied by the transformers in use. If a transformer is not used, then the current supplied is 0.

6.2.7 Capacity Limits for Generators

The current supplied by renewable generators is limited by the minimum between its capacity and the maximum power availability depending on the technology. The power factor is equal to 1.

$$0 \leq g^p_{i,t,\text{ll},w} \leq y^p_{i,t}\, \hat{G}^p_{i,t,\text{ll},w};$$

$$\forall p \in P,\ \forall i \in \Omega^p,\ \forall k \in K^p,\ \forall t \in T,\ \forall \text{ll} \in \text{LL},\ \forall w \in \Pi \tag{6.22}$$

Constraints (6.20) establish the upper and lower capacity limits for generators. Notice that wind generators upper limits depend on wind speed and PV generators upper limits depend on solar irradiation.

6.2.8 Unserved Energy

The variable associated with unserved energy, $d^U_{i,t,b}$, is defined as continuous and nonnegative. The demand is set to as the upper limit:

$$0 \leq d^U_{i,t,\text{ll},w} \leq d_{i,t,\text{ll},w};$$

$$\forall i \in \Omega^l_j,\ \forall t \in T,\ \forall \text{ll} \in \text{LL},\ \forall w \in \Pi \tag{6.23}$$

6.2.9 DG Penetration Limit

Constraints (6.24) limit the level of penetration of distributed generation as a fraction ξ of the demand. This is mathematically formulated as:

$$\sum_{p \in P} \sum_{i \in \Omega^p} g^p_{i,t,\text{ll},w} + \sum_{i \in \Omega^{\text{ST}}} (g^{\text{ST,prod}}_{i,t,\text{ll},w} - g^{\text{ST,store}}_{i,t,\text{ll},w}) \leq \varepsilon \sum_{i \in \Omega^{\text{LN}}_l} d_{i,t,\text{ll},w};$$

$$\forall \text{ll} \in \text{LL},\ \forall w \in \Pi \tag{6.24}$$

6.2.10 Capacity Limits for Storage

The power supplied or stored by storage units is limited by the minimum between their capacities and the maximum capacity (6.25) and (6.26). Finally, in (6.27), binary variables $u^{\text{ST,prod}}_{i,t,\text{ll},w}$ and $u^{\text{ST,store}}_{i,t,\text{ll},w}$ are defined to avoid producing and storing energy simultaneously.

$$\underline{G}^{\text{ST}} u^{\text{ST,prod}}_{i,t,\text{ll},w} \leq g^{\text{ST},prod}_{i,t,\text{ll},w} \leq \overline{G}^{\text{ST}} u^{\text{ST,prod}}_{i,t,\text{ll},w};$$

$$\forall i \in \Omega^i,\ \forall \text{st} \in \Omega^{\text{st}},\ \forall \text{ll} \in \text{LL},\ \forall w \in \Pi \tag{6.25}$$

$$\underline{G}^{\text{ST}} u^{\text{ST,store}}_{i,t,\text{ll},w} \leq g^{\text{ST,store}}_{i,t,\text{ll},w} \leq \overline{G}^{\text{ST}} u^{\text{ST,store}}_{i,t,\text{ll},w};$$

$$\forall i \in \Omega^i,\ \forall \text{st} \in \Omega^{\text{st}},\ \forall \text{ll} \in \text{LL},\ \forall w \in \Pi \tag{6.26}$$

$$u_{i,t,\text{ll},w}^{\text{ST,prod}} + u_{i,t,\text{ll},w}^{\text{ST,store}} \le y_{i,t}^{\text{ST}};$$

$$\forall i \in \Omega^i, \forall \text{st} \in \Omega^{\text{st}}, \forall \text{ll} \in \text{LL}, \forall w \in \Pi \tag{6.27}$$

6.2.11 Investment Constraints

It is considered that, along the whole planning horizon, it is only possible to invest in one of the candidate alternatives for each component. As per (6.27)–(6.31), a maximum of one reinforcement, replacement, or addition is allowed for each system component along the planning horizon. Constraints (6.32) guarantee that new transformers can only be added in substations that have been previously expanded or built.

$$\sum_{t \in T} \sum_{k \in K^l} x_{i,j,k,t}^l \le 1; \quad \forall l \in \{\text{NRF}, \text{NAF}\}, \forall (i,j) \in \Upsilon^l \tag{6.28}$$

$$\sum_{t \in T} x_{i,t}^{\text{SS}} \le 1; \quad \forall i \in \Omega^{\text{SS}} \tag{6.29}$$

$$\sum_{t \in T} \sum_{k \in K^{\text{NT}}} x_{i,k,t}^{\text{NT}} \le 1; \quad \forall i \in \Omega^{\text{SS}} \tag{6.30}$$

$$\sum_{t \in T} \sum_{k \in K^p} x_{i,k,t}^p \le 1; \quad \forall p \in P, \forall i \in \Omega^p \tag{6.31}$$

$$\sum_{t \in T} x_{i,t}^{\text{st}} \le 1; \quad \forall i \in \Omega^{\text{st}}, \forall t \in T \tag{6.32}$$

$$x_{i,k,t}^{\text{NT}} \le \sum_{\tau=1}^{t} x_{i,\tau}^{\text{SS}}; \quad \forall i \in \Omega^{\text{SS}}, \forall k \in K^{\text{NT}}, \forall t \in T \tag{6.33}$$

6.2.12 Utilization Constraints

The candidate assets for reinforcement, replacement, or installation can only be used once the investment is made. Mathematically, this is formulated as follows:

$$y_{i,j,k,t}^{\text{EFF}} + y_{j,i,k,t}^{\text{EFF}} = 1; \quad \forall (i,j) \in \Upsilon^{\text{EFF}}, \forall k \in K^{\text{EFF}}, \forall t \in T \tag{6.34}$$

$$y_{i,j,k,t}^l + y_{j,i,k,t}^l = \sum_{\tau=1}^{t} x_{i,j,k,\tau}^l; \quad \forall l \in \{\text{NRF}, \text{NAF}\}, \forall (i,j) \in \Upsilon^l, \forall k \in K^l, \forall t \in T$$

$$\tag{6.35}$$

$$y_{i,j,k,t}^{ERF} + y_{j,i,k,t}^{ERF} = 1 - \sum_{\tau=1}^{t} \sum_{\kappa \in K^{NRF}} x_{i,j,\kappa,\tau}^{NRF}; \quad \forall(i,j) \in \Upsilon^{ERF}, \forall k \in K^{ERF}, \forall t \in T$$

(6.36)

$$y_{i,k,t}^{NT} \leq \sum_{\tau=1}^{t} x_{i,k,\tau}^{NT}; \quad \forall i \in \Omega^{SS}, \forall k \in K^{NT}, \forall t \in T \qquad (6.37)$$

$$y_{i,t}^{p} \leq \sum_{\tau=1}^{t} x_{i,\tau}^{p}; \quad \forall i \in \Omega^{p}, \forall t \in T \qquad (6.38)$$

$$y_{i,t}^{st} \leq \sum_{\tau=1}^{t} x_{i,\tau}^{st}; \quad \forall i \in \Omega^{st}, \forall t \in T \qquad (6.39)$$

Constraints (6.34)–(6.36) model the utilization of all feeders while explicitly characterizing the direction of current flows. The utilization of new transformers is formulated in (6.37), and the utilization of installed generators and generic storage is modeled in (6.38) and (6.39).

6.2.13 Investment Limits
The total investment cost at each stage t has an upper limit that cannot be exceeded. Constraints (6.40) impose this budget limit for investment at each stage.

$$\sum_{l \in \{NRF,NAF\}} \sum_{k \in K^l} \sum_{(i,j) \in \Upsilon^l} C_{i,j,k}^{I,l} \ell_{ij} x_{i,j,k,t}^{l} + \sum_{i \in \Omega^{SS}} C_i^{I,SS} x_{i,t}^{SS} + \sum_{k \in K^{NT}} \sum_{i \in \Omega^{SS}} C_{i,k}^{I,NT} x_{i,k,t}^{NT}$$
$$+ \sum_{p \in P} \sum_{i \in \Omega^{p}} C_{ik}^{I,p} pf \overline{G}_{k}^{p} x_{i,t}^{p} + \sum_{p \in P} \sum_{i \in \Omega^{ST}} C_i^{I,ST} pf \overline{G}^{ST} x_{i,t}^{ST} \leq IB_t; \quad \forall t \in T, \forall t \in T$$

(6.40)

6.2.14 Radiality Constraints
Generally, distribution networks are radially operated regardless of their topologies. That is, distribution networks can be topologically meshed but they are operated in a radial way. This condition is modeled based on Ref. [3]:

$$\sum_{l \in L} \sum_{i \in \Omega_{j}^{l}} \sum_{k \in K^{l}} y_{i,j,k,t}^{l} = 1; \quad \forall j \in \Omega_{t}^{LN}, \forall t \in T \qquad (6.41)$$

$$\sum_{l \in L} \sum_{i \in \Omega_j^l} \sum_{k \in K^l} y_{i,j,k,t}^l \leq 1; \quad \forall j \notin \Omega_t^{\mathrm{LN}}, \forall t \in T \tag{6.42}$$

It is worth mentioning that constraints (6.41) impose nodes to have a single input flow while expression (6.42) sets a maximum of one input flow for the remaining nodes. As shown in Ref. [24], traditional radiality constraints (6.41) and (6.42) may fail to guarantee radial operation of the distribution system when DG is considered due to issues with transfer nodes and isolated generators. This difficulty is overcome by adding the following set of radiality constraints [24]:

$$\sum_{l \in L} \sum_{k \in K^l} \sum_{j \in \Omega_i^l} \left(\tilde{f}_{i,j,k,t}^l - \tilde{f}_{j,i,k,t}^l \right) = \tilde{g}_{i,t}^{\mathrm{SS}} - \tilde{D}_{i,t}; \quad \forall i \in \Omega^{\mathrm{N}}, \forall t \in T \tag{6.43}$$

$$0 \leq \tilde{f}_{i,j,k,t}^{\mathrm{EFF}} \leq n_{\mathrm{DG}}; \quad \forall i \in \Omega_j^{\mathrm{EFF}}, \forall j \in \Omega^{\mathrm{N}}, \forall k \in K^{\mathrm{EFF}}, \forall t \in T \tag{6.44}$$

$$0 \leq \tilde{f}_{i,j,k,t}^{\mathrm{ERF}} \leq \left(1 - \sum_{\tau=1}^{t} \sum_{\kappa \in K^{\mathrm{NRF}}} x_{ij\kappa\tau}^{\mathrm{NRF}} \right) n_{\mathrm{DG}}; \quad \forall (i,j) \in \Upsilon^{\mathrm{ERF}}, \forall k \in K^{\mathrm{ERF}}, \forall t \in T \tag{6.45}$$

$$0 \leq \tilde{f}_{j,i,k,t}^{\mathrm{ERF}} \leq \left(1 - \sum_{\tau=1}^{t} \sum_{\kappa \in K^{\mathrm{NRF}}} x_{ij\kappa\tau}^{\mathrm{NRF}} \right) n_{\mathrm{DG}}; \quad \forall (i,j) \in \Upsilon^{\mathrm{ERF}}, \forall k \in K^{\mathrm{ERF}}, \forall t \in T \tag{6.46}$$

$$0 \leq \tilde{f}_{i,j,k,t}^l \leq \left(\sum_{\tau=1}^{t} x_{i,j,k,\tau}^l \right) n_{\mathrm{DG}}; \quad \forall l \in \{\mathrm{NRF,NAF}\}, \forall (i,j) \in \Upsilon^l, \forall k \in K^l, \forall t \in T \tag{6.47}$$

$$0 \leq \tilde{f}_{j,i,k,t}^l \leq \left(\sum_{\tau=1}^{t} x_{i,j,k,\tau}^l \right) n_{\mathrm{DG}}; \quad \forall l \in \{\mathrm{NRF,NAF}\}, \forall (i,j) \in \Upsilon^l, \forall k \in K^l, \forall t \in T \tag{6.48}$$

$$0 \leq \tilde{g}_{i,t}^{\mathrm{SS}} \leq n_{\mathrm{DG}}; \quad \forall i \in \Omega^{\mathrm{SS}}, \forall t \in T \tag{6.49}$$

As described in Ref. [24], the existence of isolated generators is disabled through (6.43)–(6.49), which model a fictitious system with fictitious demands. According to Ref. [62], the fictitious demand at load nodes that are candidate locations for DG installation is equal to 1 pu, whereas the fictitious demand at the remaining nodes is set to 0. Mathematically:

$$\tilde{D}_{i,t} = \begin{cases} 1; & \forall i \in \Omega_t^{LN} \\ 0; & \forall i \notin \Omega_t^{LN} \end{cases} \tag{6.50}$$

Fictitious nodal demands can only be supplied by fictitious substations located at the original substation nodes, which inject fictitious power through the actual feeders. Constraints (6.43) represent the fictitious current nodal balance equations. Constraints (6.44)–(6.48) limit the fictitious flows through the feeders. Constraints (6.49) set the limits for the fictitious currents injected by the fictitious substations. The fictitious demand at load nodes that are candidate locations for DG installation is equal to 1 pu, whereas at the remaining nodes it is set to 0 (6.50).

6.2.15 Demand Response to Load Level–Changing Prices
The inclusion of DR in the problem formulation requires an adequate treatment of uncertainties in the system and load levels. In an attempt to accommodate the increasing amount of renewable energy resources (RES) in existing power systems and to provide an adequate framework for the DR impact in existing power system models, load period approach and a clustering approach have recently been applied to accurately represent load curve. DR has been introduced in the model considering elastic demand functions calibrated by load levels. The demanded energy in every load level (6.51) has been expressed as a function of elasticity, demand, and prices for the incumbent load levels included in the load-shifting horizon and average price (6.52).

$$d_{i,t,ll,w} = D_{i,t,ll,w} + \sum_{lb} \xi_{lb,ll} \frac{(D_{i,t,ll,w}\Delta_{ll} + D_{i,t,lb,w}\Delta_{lb})}{\Delta_{lb} + \Delta_{ll}} \frac{(c_{t,lb,w}^{SS} - \overline{C_{t,ll,w}^{SS}})}{\overline{C_{t,ll,w}^{SS}}} \tag{6.51}$$

$$\overline{C_{t,ll,w}^{SS}} = \frac{\sum_{lb} C_{t,lb,w}^{SS}\Delta_{lb}}{\sum_{lb}\Delta_{lb}} \tag{6.52}$$

The price elasticity assumptions determine the slope of the demand function with own-price elasticities ($\xi_{ll,ll}$) and cross-price elasticities ($\xi_{ll,lb}$) being exogenously provided, based upon values from the literature. Cross-price elasticity (also known as elasticity of substitution) refers to the consumer's reaction to the prices in other load levels and accounts for load shifting. Own-price elasticities account for immediate response to price signals in the considered load level. This study analyzes different values for own-price and cross-price elasticities. Nonconsidering DR in the expansion planning is equivalent to set own- and cross-price elasticities to zero. However, there is not yet enough experience with flexible DR programs to adequately quantify their impacts. Future research is required in this area, resulting of particular relevance to resource planning efforts, considering the expected increasing penetration of variable resources in the overall energy mix. Accurately determining the values for own- and cross-price elasticities may significantly impact the outcomes of the planning process.

6.2.16 Generic Storage Formulation

In modern power systems, storage devices have grown rapidly. Energy storage units are integrated to energy distribution systems (EDS) to meet several purposes such as real-time power demand, smoothing output power of RES, improving power system reliability and being economically efficient. The transition function of ESS is represented by (6.53). The minimum and maximum storage limits (charge) and productions (injections into the network) are defined in (6.25) and (6.26). Finally, in (6.27), binary variables, that is, $v_{i,t,ll,w}^{st,prod}$ and $v_{i,t,ll,w}^{st,store}$, are defined to avoid producing and storing energy simultaneously.

$$\sum_{n,f} \left[\Delta_{ll} \left(\eta_i^{ST,store} g_{i,t,ll,w}^{ST,store} - \left(\frac{1}{\eta_i^{ST,prod}} \right) g_{i,t,ll,w}^{ST,prod} \right) \right] = 0 \qquad (6.53)$$

6.3 LINEARIZATIONS

The optimization model for the joint generation and distribution network expansion planning presented in chapter "Energy Storage Systems Modeling" includes nonlinearities which make the optimal solution hard to obtain. The optimization model for the joint generation and distribution network expansion planning presented includes nonlinearities which make the optimal solution hard to obtain. Instead of directly addressing the original problem of mixed-integer nonlinear programming, a mixed-integer linear formulation is proposed. These nonlinearities are related to bilinear terms involving the product of continuous and binary variables, and product of continuous variables in the case of the quadratic energy losses. Both nonlinearities are recast as linear expressions by using a piecewise linear approximation for energy losses and integer algebra results for the bilinear terms. Note that mixed-integer linear programming model guarantees finite convergence to optimality while providing a measure of the distance to optimality along the solution process.

6.3.1 Energy Losses

Energy losses are modeled in (6.5) by quadratic expressions, which can be approximated by piecewise linear functions. Fig. 6.1 shows the approximation for a quadratic curve through n_ν linear chapters. A general formulation for the piecewise linearization of the quadratic curve is presented as follows (Ref. [63]):

$$\text{Current} = \sum_{\nu=1}^{n_\nu} \delta_\nu \qquad (6.54)$$

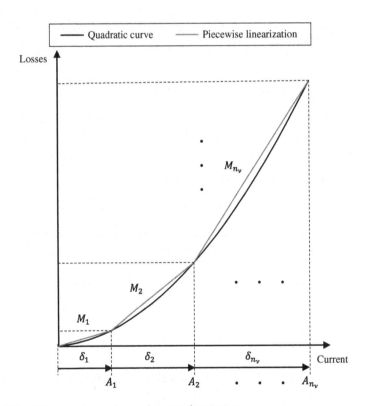

Figure 6.1 Quadratic curve of energy losses and piecewise linearization.

$$\text{Losses} = \sum_{\nu=1}^{n_\nu} M_\nu \delta_\nu \tag{6.55}$$

$$0 \le \delta_\nu \le A_\nu; \quad \nu = 1 \dots n_\nu \tag{6.56}$$

This piecewise linearization technique has been used to approximate the energy losses in existing fixed feeders, existing replaceable feeders, new replacement feeders, newly added feeders, existing transformers, and new transformers. Therefore, expression (6.5) can be linearly formulated as follows:

$$c_t^R = \sum_{w \in \Pi} \pi_w \sum_{ll \in LL} \Delta_{ll} C_{ll}^{SS} \left(\sum_{ll \in L} \sum_{k \in K^l} \sum_{(i,j) \in \Upsilon^l} \sum_{\nu=1}^{n_\nu} M_{ijk\nu}^l (\delta_{i,j,k,t,ll,\nu,w}^l + \delta_{i,j,k,t,ll,\nu,w}^l) \right.$$
$$\left. + \sum_{tr \in TR} \sum_{k \in K^{tr}} \sum_{i \in \Omega^{SS}} \sum_{\nu=1}^{n_\nu} M_{i,k,\nu}^{tr} \delta_{i,j,k,t,ll,\nu,w}^{tr} \right); \quad \forall t \in T$$

$$\tag{6.57}$$

To model the linear chapters of all energy losses considered in feeders, the following constraints are used:

$$f_{i,j,k,t,ll,w}^l = \sum_{\nu=1}^{n_\nu} \delta_{i,j,k,t,ll,\nu,w}^l; \quad \forall l \in L, \ \forall i \in \Omega_j^l, \ \forall j \in \Omega^N, \ \forall k \in K^l,$$
$$\forall t \in T, \ \forall ll \in LL, \ \forall w \in \Pi \tag{6.58}$$

$$0 \le \delta_{i,j,k,t,ll,\nu,w}^l \le A_{i,j,k,\nu}^l;$$
$$\forall l \in L, \ \forall i \in \Omega_j^l, \ \forall j \in \Omega^N, \ \forall k \in K^l, \ \forall t \in T, \ \forall ll \in LL, \ \forall w \in \Pi, \ \nu = 1 \dots n_\nu \tag{6.59}$$

Analogously, to model the linear chapters of all energy losses considered in transformers, the following constraints are used:

$$g_{i,k,t,b,w}^{tr} = \sum_{\nu=1}^{n_\nu} \delta_{i,k,t,b,\nu,w}^{tr}; \quad \forall tr \in TR, \ \forall i \in \Omega^{SS}, \ \forall k \in K^{tr}, \ \forall t \in T, \ \forall b \in B, \ \forall w \in \Pi$$

$$\tag{6.60}$$

$$0 \le \delta_{i,k,t,b,\nu,w}^{tr} \le A_{i,k,\nu}^{tr};$$
$$\forall tr \in TR, \ \forall i \in \Omega^{SS}, \ \forall k \in K^{tr}, \ \forall t \in T, \ \forall b \in B, \ \forall w \in \Pi, \ \nu = 1 \dots n_\nu \tag{6.61}$$

6.3.2 Kirchhoff's Voltage Law

Constraints (6.18) model the Kirchhoff's voltage law for all feeders in use considering existing fixed feeders, existing replaceable feeders, new replacement feeders, and newly added feeders. These constraints are only active when the corresponding line is used. This nonlinearity is modeled through binary variables and the expression associated with the Kirchhoff's voltage law. The linear formulation of these constraints is presented below:

$$-M(1 - y^l_{i,j,k,t}) \leq Z^l_{i,j,k} f^l_{i,j,k,\mathrm{ll},w} - (v_{i,t,\mathrm{ll},w} - v_{j,t,\mathrm{ll},w}) \leq M(1 - y^l_{i,j,k,t});$$

$$\forall l \in L, \ \forall i \in \Omega^l_j, \ \forall j \in \Omega^N, \ \forall k \in K^l, \ \forall \mathrm{ll} \in \mathrm{LL}, \ \forall \omega \in \Pi$$

$$(6.62)$$

In constraints (6.62) M is a large-enough positive constant and its influence is similar to constraints (6.18). If $y^l_{i,j,k,t}$ is equal to 0 there is no limit to the value of the expression in brackets in constraints (6.62), which must be between M and $-M$. In contrast, if $y^l_{i,j,k,\mathrm{ll}}$ is equal to 1, then the corresponding constraint behaves just like the corresponding constraints in (6.18).

CHAPTER 7

Case Study

Many utilities are nowadays obligated by state regulatory or legislative requirements to consider demand response (DR) as part of their resource planning process. There are several ways to incorporate DR into resource planning modeling and each has its advantages and disadvantages. What is often underappreciated about real-time pricing mechanisms is that the impacts of these programs can be very consistent and predictable due to the fact that they are based on the collective response of a large number of participants.

The Canary Islands is a Spanish Archipelago located in the northeast Atlantic, in front of the Western coast of Africa. It comprises seven islands with a total surface of 7490 km² and over 2 million inhabitants. The main islands are (from largest to smallest): Tenerife, Fuerteventura, Gran Canaria, Lanzarote, La Palma, La Gomera, and El Hierro. The archipelago also includes a number of islands and islets: La Graciosa, Alegranza, Isla de Lobos, Montaña Clara, Roque del Oeste, and Roque del Este.

The Archipelago is highly reliant on imported fossil fuels to generate electricity. Nearly 98% of the primary energy consumption is based on imported oil brought to the islands by ships (90% in terms of electricity). Volatile electricity prices create economic development challenges which can be mitigated to some extent through affordable and locally produced energy. The region has a broad range of renewable energy sources (RES), including bioenergy, wind, and photovoltaic energy. Additionally, there is a significant potential for energy efficiency improvements with regard to lighting, building, transport and distribution networks, and industrial processes (Fig. 7.1).

With the aim of analyzing the impact of DR on the joint generation and distribution expansion planning model for the La Graciosa island, the present case study present a global model of this electric power system. La Graciosa is a small island belonging to the Canary Islands located near Lanzarote, consisting of 26 nodes and 37 branches.

Joint RES and Distribution Network Expansion Planning under a Demand Response Framework.

SISTEMA ELÉCTRICO CANARIO

La Graciosa is connected to the Lanzarote–Fuerteventura electric power system. Energy consumption in La Graciosa represents less than 1% of the overall consumption in the main Lanzarote–Fuerteventura system. Therefore, the impact of generation expansion planning of La Graciosa on the main system has been dismissed. Substation prices for each load level for the connection node between the main system of Lanzarote–Fuerteventura and the system of La Graciosa are considered as input for the expansion planning model.

The base power and base voltage of the system are 1 MVA and 20 kV, respectively. A 3% interest rate is set. The lifetime of all feeders and transformers is 30 years. For simplicity, maintenance costs for all feeders are equal to €450/year. The cost of unserved energy, C^U, is €2000/MWh. Investment in DG is allowed with a penetration limit, ξ, set to 40%. The investment budget is set to €70,000 for each period. Candidate nodes for installation of wind generators are 8, 9, 10, 11, 12, 13, 15 and 16. Candidate nodes for installation of PV generators are 7, 8, 9, 12, 13, 14, 16, and 21. Candidate nodes for installation of storage units are 9, 15, 20, and 23. Upper and lower bounds for voltages at load nodes are equal to 1.05 and 0.95 pu (Fig. 7.2).

Planning models to adequately optimize generation and network investments in the distribution system of La Graciosa require a new approach to model load, wind, and irradiation levels in electric power systems. In this chapter the prices considered for the Lanzarote–Fuerteventura for the first year are presented. Results show the expected substation price for every considered block. Peak prices can be observed in winter night hours. The proposed approach is consistent with the traditional methodology where higher prices are charged to higher substation consumptions, following the rules of the marginal pricing price setting. The proposed model considers a single transmission system connected with La Graciosa distribution network. Prices considered for the proposed model are consistent with results provided by Red Eléctrica de España (REE) [64]. Simulations have been implemented on a Dell PowerEdge R910X64 with four Intel Xeon E7520 processors at 8 GHz and 32 GB of RAM using CPLEX 12 [63] under GAMS 24.0 [65]. Then, system data and results obtained are presented (Fig. 7.3).

Substation prices are integrated in the joint generation and distribution planning model for La Graciosa, modifying the overall demand profile for the island. Substation prices are not altered by the

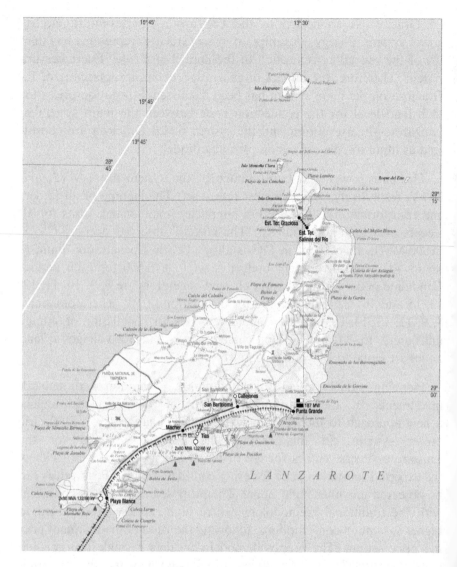

Figure 7.2 Lanzarote transmission network. Source: REE.

introduction of DR in the model formulation, being the substation always the marginal unit of the system.

7.1 LA GRACIOSA CASE STUDY

The distribution network analyzed consists of 23 load nodes, 3 substation nodes, and 37 branches. Demand nodes, existing substations,

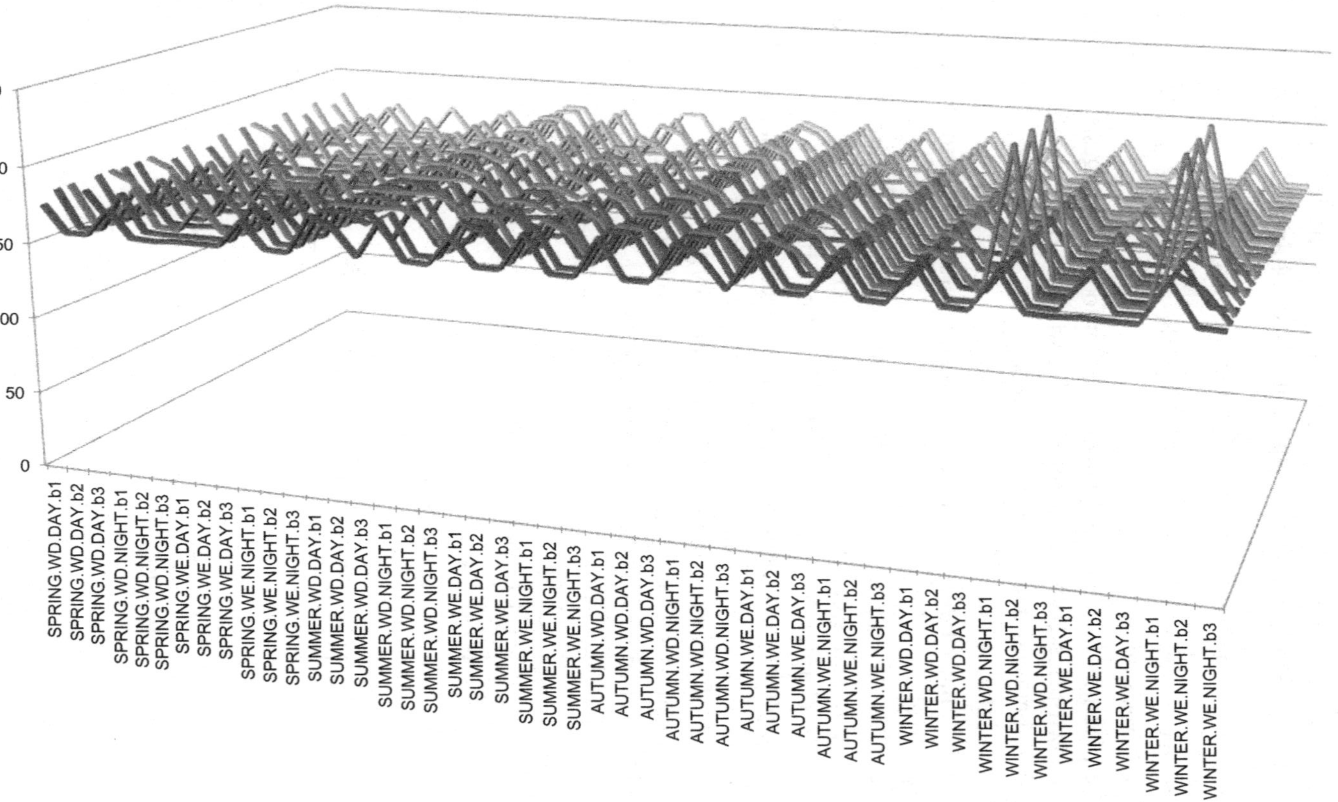

Figure 7.3 Substation prices for La Graciosa.

The x-axis labels, reading in order, are:

SPRING.WD.DAY.b1
SPRING.WD.DAY.b2
SPRING.WD.DAY.b3
SPRING.WD.NIGHT.b1
SPRING.WD.NIGHT.b2
SPRING.WD.NIGHT.b3
SPRING.WE.DAY.b1
SPRING.WE.DAY.b2
SPRING.WE.DAY.b3
SPRING.WE.NIGHT.b1
SPRING.WE.NIGHT.b2
SPRING.WE.NIGHT.b3
SUMMER.WD.DAY.b1
SUMMER.WD.DAY.b2
SUMMER.WD.DAY.b3
SUMMER.WD.NIGHT.b1
SUMMER.WD.NIGHT.b2
SUMMER.WD.NIGHT.b3
SUMMER.WE.DAY.b1
SUMMER.WE.DAY.b2
SUMMER.WE.DAY.b3
SUMMER.WE.NIGHT.b1
SUMMER.WE.NIGHT.b2
SUMMER.WE.NIGHT.b3
AUTUMN.WD.DAY.b1
AUTUMN.WD.DAY.b2
AUTUMN.WD.DAY.b3
AUTUMN.WD.NIGHT.b1
AUTUMN.WD.NIGHT.b2
AUTUMN.WD.NIGHT.b3
AUTUMN.WE.DAY.b1
AUTUMN.WE.DAY.b2
AUTUMN.WE.DAY.b3
AUTUMN.WE.NIGHT.b1
AUTUMN.WE.NIGHT.b2
AUTUMN.WE.NIGHT.b3
WINTER.WD.DAY.b1
WINTER.WD.DAY.b2
WINTER.WD.DAY.b3
WINTER.WD.NIGHT.b1
WINTER.WD.NIGHT.b2
WINTER.WD.NIGHT.b3
WINTER.WE.DAY.b1
WINTER.WE.DAY.b2
WINTER.WE.DAY.b3
WINTER.WE.NIGHT.b1
WINTER.WE.NIGHT.b2
WINTER.WE.NIGHT.b3

The y-axis (vertical) is scaled: 0, 50, 100, 150, 200, 250.

candidate substation, existing feeders, replaceable feeders, and candidates for the installation of new feeders are shown in the schematic distribution network topology presented in Figs. 7.4 and 7.5.

Data associated with the optimization problem are:

- The base power and base voltage of the system are 1 MVA and 20 kV, respectively.
- The planning horizon is 3 years divided into yearly stages.
- The own-price elasticity is set to 0.1 and cross-price elasticity is set to 0.02 between day and night periods.
- A 10% interest rate is set.
- The lifetime of all feeders and transformers is 25 and 15 years, respectively.
- The costs of energy supplied by all substations, C_b^{SS} are obtained from the solution of the Lanzarote–Fuerteventura problem as stated in chapter "Optimization Problem Formulation."
- For simplicity, maintenance costs for all feeders are equal to €450/year.
- The cost of unserved energy, C^U, is €2000/MVAh.
- Investment in DG (including storage) is allowed with a penetration limit, ξ, set to 25% of the peak demand.

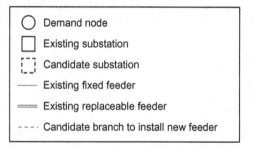

Figure 7.4 List of solution symbols.

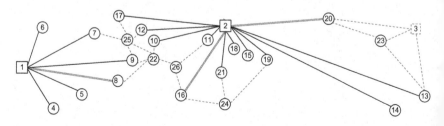

Figure 7.5 One-line diagram of the distribution network.

- The investment budget is set to €140,000.
- Candidate nodes for installation of wind generators are 8, 9, 10, 11, 13, and 16.
- Candidate nodes for installation of PV generators are 7, 12, 13, 14, 16, and 21.
- Candidate nodes for installation of Hybrid Storage Units are 9, 15, 20 and 23.
- Upper and lower bounds for voltages at load nodes are equal to 1.05 and 0.95 pu, respectively.
- Voltages at substation nodes are set to 1.05 pu.
- A three-block piecewise linearization is used to approximate energy losses.

Demand levels for each load block are presented in Fig. 7.6. Summer demand peaks can be observed in the figure, where demand is nearly 50% higher than during nontouristic periods. The winter peak corresponding to New Year's period has also been captured by the proposed methodology.

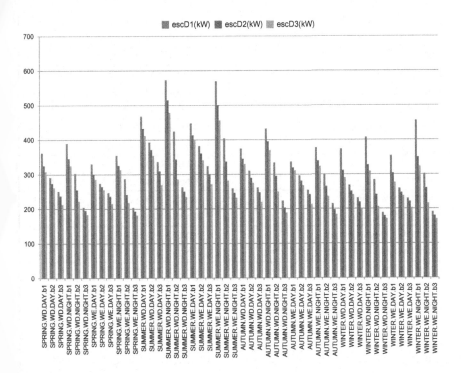

Figure 7.6 Demand levels for La Graciosa.

Table 7.1 shows the duration of each considered block computed in hours. The total number of hours of the first block is lower than hours included in blocks 2 and 3, in order to accurately include peak demand in the model. For simplicity reasons, blocks 2 and 3 have been defined with the same size.

Wind speed levels for each load block are presented in Fig. 7.7. It can be observed how winter blocks correspond to higher wind speeds, as could be expected. Additionally, no sensible differences can be observed when comparing day and night values.

Irradiation levels for each load block are presented in Fig. 7.8. It can be observed how spring blocks correspond to higher irradiation data.

Table 7.1 Hours Included in Each Considered Block			
Block	Hours	Block	Hours
SPRING.WD.DAY.b1	48	SUMMER.WD.DAY.b1	54
SPRING.WD.DAY.b2	134	SUMMER.WD.DAY.b2	134
SPRING.WD.DAY.b3	134	SUMMER.WD.DAY.b3	134
SPRING.WD.NIGHT.b1	34	SUMMER.WD.NIGHT.b1	39
SPRING.WD.NIGHT.b2	90	SUMMER.WD.NIGHT.b2	92
SPRING.WD.NIGHT.b3	90	SUMMER.WD.NIGHT.b3	92
SPRING.WE.DAY.b1	22	SUMMER.WE.DAY.b1	23
SPRING.WE.DAY.b2	55	SUMMER.WE.DAY.b2	50
SPRING.WE.DAY.b3	55	SUMMER.WE.DAY.b3	50
SPRING.WE.NIGHT.b1	17	SUMMER.WE.NIGHT.b1	14
SPRING.WE.NIGHT.b2	37	SUMMER.WE.NIGHT.b2	37
SPRING.WE.NIGHT.b3	37	SUMMER.WE.NIGHT.b3	37
AUTUMN.WD.DAY.b1	31	WINTER.WD.DAY.b1	40
AUTUMN.WD.DAY.b2	100	WINTER.WD.DAY.b2	123
AUTUMN.WE.DAY.b3	100	WINTER.WD.DAY.b3	123
AUTUMN.WD.NIGHT.b1	26	WINTER.WD.NIGHT.b1	65
AUTUMN.WD.NIGHT.b2	100	WINTER.WD.NIGHT.b2	100
AUTUMN.WD.NIGHT.b3	100	WINTER.WE.NIGHT.b3	100
AUTUMN.WE.DAY.b1	14	WINTER.WE.DAY.b1	20
AUTUMN.WE.DAY.b2	38	WINTER.WE.DAY.b2	50
AUTUMN.WE.DAY.b3	38	WINTER.WE.DAY.b3	50
AUTUMN.WE.NIGHT.b1	19	WINTER.WE.NIGHT.b1	18
AUTUMN.WE.NIGHT.b2	33	WINTER.WE.NIGHT.b2	47
AUTUMN.WE.NIGHT.b3	33	WINTER.WE.NIGHT.b3	47

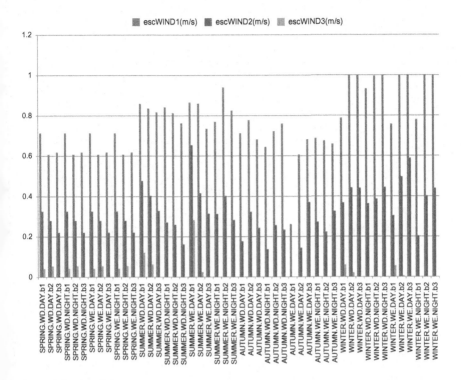

Figure 7.7 Wind speed.

Data regarding branch lengths are listed in Table 7.2.

Table 7.3 shows data for existing fixed feeders. These data comprise maximum current flow, impedance, and resistance.

Table 7.4 shows data for candidate feeders in branches subject to replacement.

Table 7.5 shows data for candidate feeders in nonexisting branches.

Table 7.6 shows data for existing transformers, candidate transformers to install, and the cost of expanding or building substations.

The economic and technical features of candidate DG units are presented in Table 7.7. For simplicity reasons, only one alternative has been considered for wind and photovoltaic generation units. Data corresponding to the considered storage alternatives are presented in Table 7.8.

The power curve of the wind generator is presented in Fig. 7.9.

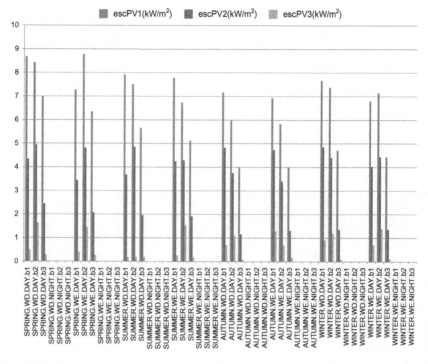

Figure 7.8 Irradiation data.

Table 7.2 Branch Lengths (m)

Branch		ℓ_{ij}	Branch		ℓ_{ij}	Branch		ℓ_{ij}
i	j		i	j		i	j	
1	4	200	2	17	230	10	22	050
1	5	185	2	18	030	11	26	112
1	6	090	2	19	160	12	25	043
1	7	190	2	20	240	13	23	270
1	8	260	2	21	150	16	24	155
1	9	250	3	13	220	16	26	090
2	10	200	3	20	290	17	25	66
2	11	020	3	23	120	19	24	220
2	12	220	7	25	053	20	23	170
2	13	500	8	22	082	21	24	093
2	14	550	9	22	064	22	25	085
2	15	050	9	25	046	22	26	104
2	16	220						

Table 7.3 Data for Existing Fixed Feeders

$\overline{F}_{ij}^{\text{EFF}}$ (MVA)	Z_{ij}^{EFF} (Ω/km)	R_{ij}^{EFF} (Ω/km)
0.1	0.5568	0.4070

Table 7.4 Data for Candidate Replacement Conductors

Alternative 1				Alternative 2			
$\overline{F}_{ij1}^{\text{NRF}}$ (MVA)	Z_{ij1}^{NRF} (Ω/km)	R_{ij1}^{NRF} (Ω/km)	$C_{ij1}^{\text{I,NRF}}$ (10^3 €/km)	$\overline{F}_{ij2}^{\text{NRF}}$ (MVA)	Z_{ij2}^{NRF} (Ω/km)	R_{ij2}^{NRF} (Ω/km)	$C_{ij2}^{\text{I,NRF}}$ (10^3 €/km)
0.3	0.2326	0.2100	30.2	0.5	0.1920	0.1227	35.3

Table 7.5 Data for Candidate Conductors in Nonexisting Branches

Alternative 1				Alternative 2			
$\overline{F}_{ij1}^{\text{NAF}}$ (MVA)	Z_{ij1}^{NAF} (Ω/km)	R_{ij1}^{NAF} (Ω/km)	$C_{ij1}^{\text{I,NRF}}$ (10^3 €/km)	$\overline{F}_{ij1}^{\text{NAF}}$ (MVA)	Z_{ij1}^{NAF} (Ω/km)	R_{ij2}^{NAF} (Ω/km)	$C_{ij2}^{\text{I,NRF}}$ (10^3 €/km)
0.1	0.5568	0.4070	32.3	0.3	0.2326	0.2100	38.7

Table 7.6 Data for Transformers

Node i	Existing Transformer			$C_i^{\text{I,SS}}$ (k€)	Alternative 1				Alternative 2			
	$\overline{G}_{i1}^{\text{ET}}$ (MVA)	R_{i1}^{ET} (Ω)	$C_{i1}^{\text{M,ET}}$ (k€)		$\overline{G}_{i1}^{\text{NT}}$ (MVA)	R_{i1}^{NT} (Ω)	$C_{i1}^{\text{M,NT}}$ (k€)	$C_{i1}^{\text{I,NT}}$ (k€)	$\overline{G}_{i2}^{\text{NT}}$ (MVA)	R_{i2}^{NT} (Ω)	$C_{i2}^{\text{M,NT}}$ (k€)	$C_{i2}^{\text{I,NT}}$ (k€)
1	0.4	0.25	0.1	4	0.6	0.16	0.2	25	0.8	0.13	0.3	32
2	0.4	0.25	0.1	4	0.6	0.16	0.2	25	0.8	0.13	0.3	32
3	–	–	–	6	0.6	0.16	0.2	25	0.8	0.13	0.3	32

Table 7.7 Data for Candidate DG Units

Alternative k	$\overline{G}_k^{\text{W}}$ (MVA)	$C_k^{\text{I,W}}$ (€/MVA)	$C_k^{\text{E,W}}$ (€/MVAh)	$\overline{G}_k^{\Theta}$ (MVA)	$C_k^{\text{I},\Theta}$ (€/MVA)	$C_k^{\text{E},\Theta}$ (€/MVAh)
1	0.0175	1010	5	0.024	500	4

Table 7.8 Data for Candidate Storage Units

Alternative k	$\overline{G}^{\text{st}}$ (MVA)	$C^{\text{I,st}}$ (€/MVA)	$C^{\text{st,prod}}$ (€/MVAh)	$C_k^{\text{st,store}}$ (€/MVAh)
1	0.02	2000	1	1

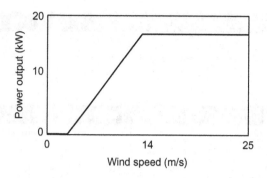

Figure 7.9 Power output for wind generator alternative 1.

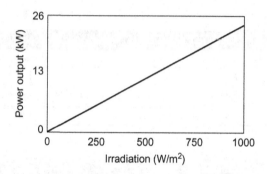

Figure 7.10 Power output for photovoltaic generator alternative 1.

The power curve of the photovoltaic generator is presented in Fig. 7.10.

Table 7.9 presents the number of customers, Nq_{it}, in every node at each stage.

7.2 RESULTS WITHOUT DR AND HYBRID STORAGE

In this section, the results obtained by the expansion planning algorithm for the case study are presented. Then, a comparative analysis is done. Fig. 7.11 shows the symbols used to represent the topology of the solutions obtained (Figs. 7.12–7.14), where the installation stage is represented by the number of astherics in the figure.

Numerical results presented in Table 7.10 represent how to expand the generation and distribution network adding renewable generation and new assets (lines and substations) so that the current and future energy supply in islands is served at a minimum cost and with the quality required.

Node	Stage			Node	Stage		
	1	**2**	**3**		**1**	**2**	**3**
4	35	37	38	16	104	106	107
5	30	32	37	17	026	027	028
6	17	21	25	18	009	012	014
7	42	44	46	19	006	007	009
8	52	54	55	20	001	002	004
9	03	08	09	21	004	005	005
10	69	72	74	22	096	097	097
11	21	23	26	23	052	054	057
12	26	29	30	24	000	010	010
13	65	68	70	25	000	026	030
14	50	51	55	26	000	000	038
15	32	34	35				

Table 7.9 Number of Customers per Node

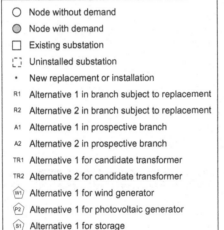

Figure 7.11 List of solution symbols.

The objective function is to maximize the net social welfare of the system and represents the total payment of the consumers minus the present value of the total cost, which consists of six cost terms related to investment, maintenance, production, losses, unserved energy, and storage. As described in Section 7.1, when inelastic demand is considered, the maximization of the net social welfare is equivalent to the minimization of the considered costs. Payment of the demand for each

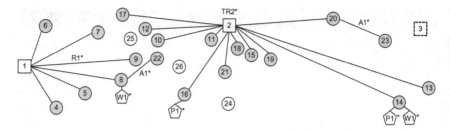

Figure 7.12 Expansion planning for the first year in the base case.

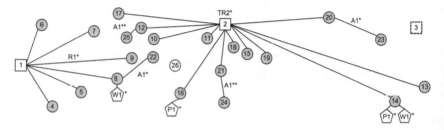

Figure 7.13 Expansion planning for the second year in the base case.

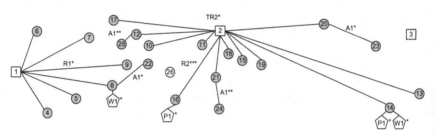

Figure 7.14 Expansion planning for the third year in the base case.

Table 7.10 Summary of Operating and Investment Costs (Thousands of €)				
Costs	Year 1	Year 2	Year 3	Total
Investment	108.066	4.088	2.780	114.934
Maintenance	10.973	10.720	13.421	132.608
Energy purchased	259.212	262.353	358.124	3481.264
Losses	0.025	0.027	0.041	0.387
Unserved energy	0.000	0.000	0.000	0.000
Total	378.275	277.187	374.665	3729.193

considered year is presented in Table 7.11. Please note that all presented costs are actualized to the first year. According to (6.1), investment costs are amortized in annual payments during the lifetime of the installed equipment, considering that once the component is operated

Table 7.11 Payment of the Demand (Thousands of €)				
	Year 1	Year 2	Year 3	Total
Payment	403.665	448.745	505.126	4912.426

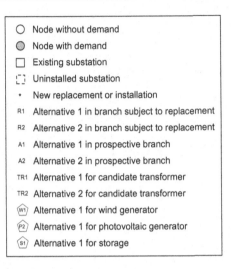

○   Node without demand

◉   Node with demand

□   Existing substation

⌐⌐   Uninstalled substation

*   New replacement or installation

R1   Alternative 1 in branch subject to replacement

R2   Alternative 2 in branch subject to replacement

A1   Alternative 1 in prospective branch

A2   Alternative 2 in prospective branch

TR1   Alternative 1 for candidate transformer

TR2   Alternative 2 for candidate transformer

(W1)   Alternative 1 for wind generator

(P2)   Alternative 1 for photovoltaic generator

(S1)   Alternative 1 for storage

Figure 7.15 List of solution symbols.

during a time equal to its lifetime, there is a reinvestment in identical equipment, so infinite annual updated payments are used. The remaining costs related to operation are updated and these costs are kept indefinitely, taking into account an infinite series of annual payments.

7.3 RESULTS WITH DR

In this section, the results obtained by the expansion planning algorithm for the case study with DR are presented. Then, a comparative analysis is done. Fig. 7.15 shows the symbols used to represent the topology of the solutions obtained.

The study demonstrates that the effect of the inclusion of DR on the model formulation is the deferment of the capacity enhancement driven by natural demand growth. Investment in renewable technologies in the first year has been reduced. During the second and third year, deferred investments take place, increasing the overall investment when comparing these results with those obtained for the previous case. Figs. 7.16—7.18 show graphically the above-mentioned affects. The model results in a total payment of the demand lower than in the

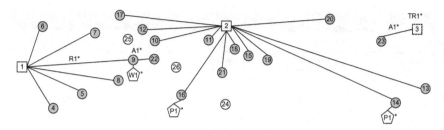

Figure 7.16 Expansion planning for the first year with DR.

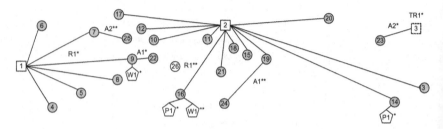

Figure 7.17 Expansion planning for the second year with DR.

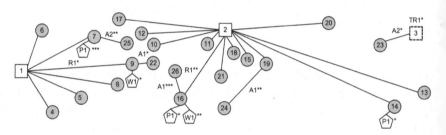

Figure 7.18 Expansion planning for the third year with DR.

first case, where DR has not been considered. The overall purchased energy costs are reduced over the three considered periods, as a consequence of load shifting to lower price load levels.

Total costs for the generation and distribution company are reduced over the considered period, as well as total payment of the consumers. Consumers should have the opportunity to receive and respond to prices or signals that reflect real-time conditions. The long-term benefit allowing demand to respond to time-varying price conditions is a flatter load shape, which should reduce the need for peaking capacity and, in turn, reduce emissions and costs through the more effective and efficient use of the grid. Both, generation and distribution companies and final consumers are benefited in long-term planning from the exposure of consumers to real-time prices.

Table 7.12 Summary of Operating and Investment Costs (Thousands of €)				
Costs	Year 1	Year 2	Year 3	Total
Investment	50.856	19.442	35.605	105.903
Maintenance	10.069	10.554	13.321	130.712
Energy purchased	257.321	248.836	339.139	3308.963
Losses	0.030	0.029	0.033	0.334
Unserved energy	0.000	0.000	0.000	0.000
Total	318.276	278.861	388.099	3545.911

Table 7.13 Payment of the Demand (Thousands of €)				
	Year 1	Year 2	Year 3	Total
Payment	383.571	426.411	479.985	4667.929

Benefits of the integration of DR in the generation and distribution expansion planning are supported by the numerical results presented in Table 7.12, such as investment deferment and reduction of the payment of the demand. Total costs afforded by generation and distribution companies are reduced by roughly 5% with the integration of DR, while overall payment over the considered time horizon is reduced by 5% (Table 7.13).

7.4 RESULTS WITH HYBRID STORAGE

In this section, the results obtained by the expansion planning algorithm for the case study with ESS are presented. Then, a comparative analysis is done. Fig. 7.19 shows the symbols used to represent the topology of the solutions obtained.

The study demonstrates that the effect of the inclusion of ESS in the model formulation accelerates the investment in renewable technologies (including ESS). Investment in storage during the first year takes place, significantly increasing the overall investment. During the second and third years investment decisions remain the same as for the previous case (Figs. 7.20–7.22). The model results in a total payment for the demand equal to the one in the base case. The overall purchased energy costs are reduced over the three considered periods, as a consequence of the management of the installed storage systems.

The total costs for the generation and distribution company are reduced over the considered period, since storage systems are

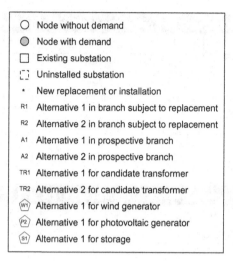

○   Node without demand
◉   Node with demand
□   Existing substation
⌐⌐   Uninstalled substation
∗   New replacement or installation
R1 Alternative 1 in branch subject to replacement
R2 Alternative 2 in branch subject to replacement
A1 Alternative 1 in prospective branch
A2 Alternative 2 in prospective branch
TR1 Alternative 1 for candidate transformer
TR2 Alternative 2 for candidate transformer
Ⓦ1 Alternative 1 for wind generator
Ⓟ2 Alternative 1 for photovoltaic generator
Ⓢ1 Alternative 1 for storage

Figure 7.19 List of solution symbols.

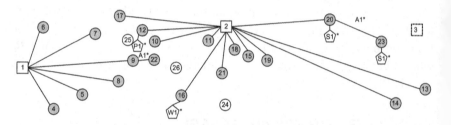

Figure 7.20 Expansion planning for the first year with ESS.

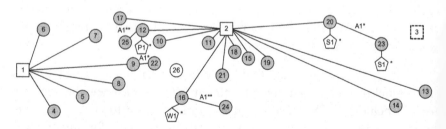

Figure 7.21 Expansion planning for the second year with ESS.

controlled by the company. The operational analysis of the considered system provides a hint that energy storage investment is a feasible solution with increased wind and solar penetration.

Benefits of the integration of ESS in generation and distribution expansion planning are supported by the numerical results presented in

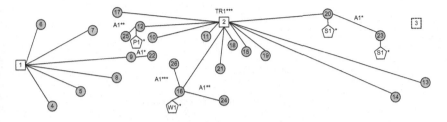

Figure 7.22 Expansion planning for the third year with ESS.

Table 7.14 Summary of Operating and Investment Costs (Thousands of €)

Costs	Year 1	Year 2	Year 3	Total
Investment	124.503	21.265	35.605	181.374
Maintenance	13.341	13.530	16.921	166.712
Energy purchased	270.076	253.310	344.133	3367.459
Losses	0.032	0.030	0.036	0.363
Unserved energy	0.000	0.000	0.000	0.000
Total	407.952	288.135	396.695	3715.908

Table 7.15 Payment of the Demand (Thousands of €)

	Year 1	Year 2	Year 3	Total
Payment	403.665	448.745	505.126	4912.426

Table 7.14. Total costs afforded by generation and distribution companies are reduced by roughly 0.5% with the integration of ESS in the expansion planning model formulation (Table 7.15).

7.5 RESULTS WITH DR AND HYBRID STORAGE

In this chapter, the results obtained by the expansion planning algorithm for the case study are presented. Then, a comparative analysis is done. Fig. 7.23 shows the symbols used to represent the topology of the solutions obtained (Figs. 7.24–7.26).

The introduction of hybrid storage in the model formulation accelerates the investment in renewable technologies, allowing a higher penetration and accounting for wind and PV imbalances. Results in Table 7.16 show how all investment decisions are made during the first 2 years, being the total investment in the third year nearly zero. Additionally, the overall investment over the considered horizon is

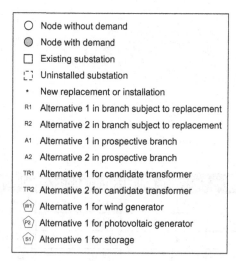

○	Node without demand
◉	Node with demand
☐	Existing substation
⸣⸢	Uninstalled substation
•	New replacement or installation
R1	Alternative 1 in branch subject to replacement
R2	Alternative 2 in branch subject to replacement
A1	Alternative 1 in prospective branch
A2	Alternative 2 in prospective branch
TR1	Alternative 1 for candidate transformer
TR2	Alternative 2 for candidate transformer
Ⓦ1	Alternative 1 for wind generator
Ⓟ2	Alternative 1 for photovoltaic generator
Ⓢ1	Alternative 1 for storage

Figure 7.23 List of solution symbols.

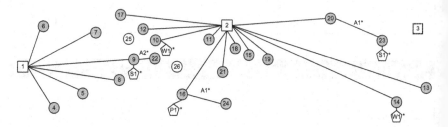

Figure 7.24 Expansion planning for the first year with DR and ESS.

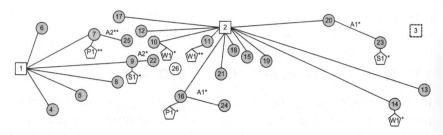

Figure 7.25 Expansion planning for the second year with DR and ESS.

higher than in the previous cases, since different storage units are foreseen for the first 2 years, together with wind and PV generation units. It is worth mentioning how the integration of ESS results in a reduction of network investment cost. Unlike the previous cases, no investment in substation is required. The model results in a total payment of the demand lower than in the first case, where DR and hybrid storage

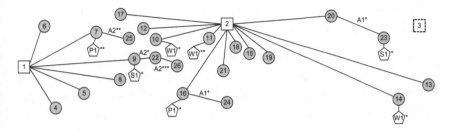

Figure 7.26 Expansion planning for the third year with DR and ESS.

Table 7.16 Summary of Operating and Investment Costs (Thousands of €)				
Costs	Year 1	Year 2	Year 3	Total
Investment	122.679	63.804	2.780	220.836
Maintenance	13.246	15.228	18.876	184.475
Energy purchased	238.255	232.812	319.984	3115.565
Losses	0.025	0.027	0.042	0.396
Unserved energy	0.000	0.000	0.000	0.000
Total	374.205	311.871	341.682	3521.273

Table 7.17 Payment of the Demand (thousands of €)				
	Year 1	Year 2	Year 3	Total
Payment	383.574	426.410	479.985	4667.929

have not been considered. Distribution and generation companies manage storage units to increase their expected benefits, accommodating a higher renewable penetration. It can be observed how total costs for the generation and distribution companies are reduced by 5.6% when compared with the results obtained for the base case. Additionally the payment of the consumers over the consumers over the consumers time horizon is reduced by 5% (Table 7.17).

7.6 IMPACT OF DR AND HYBRID STORAGE IN GENERATION AND DISTRIBUTION EXPANSION PLANNING

DR and hybrid storage are becoming a promising subject in operation and planning of electrical power systems. However, the impact of both technologies on joint generation and distribution expansion planning has not been fully analyzed yet. This study investigates the effect of these two technologies on the expansion planning presenting relevant results for agents responsible of system planning.

Responsive demand has the potential to play an important role in more flexible and smarter power systems. Its relevance in short-term operation of electric power systems has been extensively investigated. When considering the long-term planning, demand responsiveness has an impact on reduction or deferment in required expansion planning. DR can even substitute generation and distribution network expansion, as shown in this study, where generation and network investments have been deferred.

By integrating demand elasticity, consumers are able to adjust their consumption in response to real-time price signals. With demand responsive consumers, lower and higher net demand levels result in lower and higher electricity prices, respectively, in a particular load level. As presented in the study, the inclusion of DR in the formulation of the problem reduces the payment of consumers, partially shifting their consumption to low price levels.

The results show that the demand responsiveness results in postponing installation of new units and reduces the capacity that should be installed. Additionally this study investigates the costs and benefits of hybrid storage deployment and the reduction of network investment cost by deploying hybrid storage providing numerical results. Numerical results presented in Sections 7.2–7.4 confirm the potential of DR and hybrid storage to benefit consumers, reducing the payment over the considered time horizon. Additionally, it has been shown how network reinforcement and generation investments have been deferred, reducing overall costs for generation and distribution companies.

The integration of hybrid storage technologies in the overall energy mix results in a higher penetration of renewable technologies, while reducing at the same time the costs for generation and distribution companies (including substation payment). Payment of the consumers is not altered by hybrid storage, since the marginal technology remains the substation. Storage is used by planners to reduce their costs, while speeding up the integration of wind and PV. Additionally, network investment is reduced, since no reinforcement of substations is required.

Tables 7.18 and 7.19 present the present values of operational and investment costs and present value of total payment of the demand.

Table 7.18 Summary of Total Operating and Investment Costs (Thousands of €)				
Costs	Base Case	DR	ESS	DR and ESS
Investment	114.934	105.903	181.374	220.836
Maintenance	132.608	130.712	166.712	184.475
Energy purchased	3481.264	3308.963	3367.459	3115.565
Losses	0.387	0.334	0.363	0.396
Unserved energy	0.000	0.000	0.000	0.000
Total	3729.193	3545.911	3715.908	3521.273

Table 7.19 Total Payment of the Demand (Thousands of €)				
	Base Case	DR	ESS	DR and ESS
Payment	4912.426	4667.930	4912.426	4667.929

DR impacts costs and payment reducing all considered items. ESS accelerates the integration of renewable technologies, increasing investment costs and therefore reducing energy purchased from the substation.

7.7 COST–BENEFIT ANALYSIS

Generation and distribution expansion planning have been completely modified as a consequence of the restructuration and liberalization of electric power systems, since the final decision on the adequacy of investing in new generating facilities relies on investors, based on individual cost–benefit analysis (CBA). Due to its singularities (small not interconnected systems, with reduced competitiveness and no market infrastructure), the situation in isolated power systems as the one considered here is not completely the same. However, a cost–benefit analysis can be used to measure the impact of DR and ESS in cost and revenue stream for both consumers and generation and distribution companies.

Each situation is studied from the perspective of its estimated net profit over a period of several years, which depends for the considered study on the elasticity of the demand, the integration of ESS in the formulation of the expansion planning and the combination of both effect. Financial risk assessment plays a crucial role in such reviews because it is the key to obtaining adequate financing.

CBA, sometimes called benefit–cost analysis, is a process for calculating and comparing benefits and costs of a project, decision or government policy. In this study, this has been used to calculate and compare cost and benefits of the considered case studies over the defined time horizon. The value of the benefits derived from optimal investment planning is subtracted from the total costs of undertaking this investment planning. The effects of the benefits are calculated during the payback period by getting their present values and comparing it with the present values of the costs (6.1)–(6.7). Therefore, CBA provides information about the profitability of an investment combining initial investment costs and arising benefits in a time horizon that is preset. Benefits for the consumers are calculated as the difference between the payment in the considered case study and the base case. Benefits of the integration of DR for the consumers in the model formulation arise from the reduction of its payment, and the base case can be, therefore, considered as baseline.

Present values for costs and benefits for both generation and distribution companies and consumers are presented in Table 7.20. Results present CBA for all considered case studies:

- Base case where neither DR nor ESS is considered in generation and distribution expansion planning
- Integration of DR in generation and distribution expansion planning
- Integration of ESS in generation and distribution expansion planning
- Integration of both DR and ESS in generation and distribution expansion planning

DR can be defined as an incentive payment to reduce electricity consumption in times of high energy prices, increasing the electricity consumption at times of low prices. Any consumption shifting from

Table 7.20 Cost–Benefit Analysis (Thousands of €)				
	Base Case	DR	ESS	DR and ESS
Benefits for Generation & Distribution Company	1183.233	1122.018	1196.519	1146.657
Costs for Generation & Distribution Company	3729.193	3545.911	3715.908	3521.273
CBA for Generation & Distribution Company	31.73%	31.64%	32.20%	32.56%
Benefits for Consumers	0.000	244.497	0.000	244.497
Costs for Consumers	4912.426	4667.930	4912.426	4667.929
CBA for Consumers	0.00%	5.24%	0.00%	5.24%

high energy prices to low energy prices will reduce benefits for generation and distribution companies, while increasing the benefits for consumers. Additionally, the introduction of DR in the problem formulation results in lower costs for generation and the distribution company, reducing the overall investment from €3729.193 to €3545.911. However, CBA for the generation and distribution company is reduced from 31.73% to 31.64%.

Energy storage can benefit utilities allowing transmission and distribution upgrade deferrals and can speed up the integration of renewable technologies. In the present formulation, ESS are controlled by generation companies, and, therefore, its introduction increases the benefits for generation and the distribution company. CBA is increased from 31.73% to 32.20% when considering ESS in the expansion planning model. Costs for the company are reduced (from €3729.193 to €3715.908) over the considered time horizon, while benefits are increased (from €1183.233 to €1196.519).

The introduction of DR and ESS in the model formulation results in the most favorable scenario for consumers and producers. Costs for the company are reduced (from €3729.193 to €3521.273) over the considered time horizon, while benefits associated to the investment are reduced with factor 5 when compared to the costs (from €1183.233 to €1146.657). This situation results in an increase in CBA from 31.73% to 32.56%. Consumers benefit as well from the integration of DR, reducing their expected payment by €244.497.

The presented results outline that an adequate expansion planning requires the integration of DR and hybrid storage in the planning process, since some overinvestments may be averted. However, comprehensive investigation of all impacts of DR on investment planning needs more research.

CHAPTER 8

Summary and Conclusions

8.1 SUMMARY

In this work, a generation and distribution network expansion planning algorithm has been presented from a centralized viewpoint. The presented model includes DR and hybrid storage. Finally, an investment decision has been adopted through a comparative analysis. The installation of feeders, transformers, substations, and generators has been considered and adequately described in the expansion planning algorithm. This allows a distribution company to obtain the optimal strategy to meet a rising demand. A complete planning framework is considered, including the one-time investment costs as well as the long-term operational and maintenance costs.

DR has been introduced in the model considering elastic demand functions calibrated by load levels. Demanded energy in every load level has been expressed as a function of the elasticity, demand, and prices for the incumbent load levels included in the load-shifting horizon and average price.

The optimization model has been mathematically formulated as a mathematical programming problem. The objective function consists of the minimization of costs associated with investment, maintenance, supplied energy, energy losses, and nonserved energy. This minimization of the objective function is subject to technical, economic, and power balance constraints. The expansion planning algorithm, which includes the previous optimization problem, has been depicted. The modeling of uncertainty in a distribution system for each single load block is based on the load, wind, and solar irradiation curves. Hourly historical demand data are arranged from higher to lower values keeping the correlation between the different hourly data of wind and PV productions. The load duration curve is approximated using demand blocks. The corresponding scenario-based deterministic equivalent is formulated as a mixed-integer nonlinear program. The subsequent use of some well-known linearization schemes yields a mixed-integer linear program suitable for efficient off-the-shelf software.

It can be outlined that an adequate expansion planning requires the integration of DR and ESS in the planning process, since some overinvestments may be averted. Operation and investment costs, as well as payment of the demand for the considered models outline the potential of DR to effectively reduce payments of the demand in the long term by approximately 5%. Additionally the impact of ESS on the generation and distribution expansion planning has been exhaustively investigated, showing the potential benefits for the central planner.

The methodology proposed has been illustrated with a case study composed of a distribution system of 26 nodes and 37 branches, where the planning horizon has been 3 years divided into yearly stages.

8.2 CONCLUSIONS

This study introduces an algorithm to decide the joint expansion planning of distributed generation and distribution network. The outcomes of the model are the location and size of the new generation assets to be installed when fixed and variable costs are known. The model introduces other issues relevant to planning in insular distribution systems, including DR and hybrid storage. The integration of price-dependent resources, such as DR or hybrid storage, calls for a review in the traditional methodology to approximate the demand curve by load levels, increasing the number of load blocks. A more detailed representation of the demand, wind, and irradiation curves better captures chronological information, resulting in a more accurate representation of system outcomes. A total of 48 periods have been considered in the model formulation. This representation is superior to traditional monotone demand curves.

DR was originally developed by electric utilities in order to increase flexibility of the demand side by temporarily shifting or reducing peak energy demand, thereby avoiding costly energy procurements and capacity investments for a small number of hours of need. Expansion of variable renewable capacity, particularly wind and solar energy, increases the need for flexible resources, particularly those with an element of storage, that are capable of maintaining the balance between generation and load under normal conditions.

When DR is integrated in the expansion planning model, the traditional minimization of costs formulation partially disregard the benefits consumers receive from the modification of their electricity consumption. Social welfare maximization that considers both generation and demand replaces cost minimization. The study demonstrates that the effect of the inclusion of DR in the model formulation is the deferment of the capacity enhancement driven by natural demand growth. The model results in the reduction of total payment of the demand and total costs for the generation and distribution company as a consequence of load shifting to lower prices load levels.

This study investigates the costs and benefits of ESS deployment. Storage is becoming a major issue with the increasing penetration of renewable but decentralized energy sources in power networks. Some relevant conclusions are drawn from the test results. There is obviously a cost associated to storing energy, but outcomes of the presented study show how this alternative may result cost effective with the considered data. Reduction of network investment cost can be achieved by deploying ESS. Additionally the integration of hybrid storage technologies in the overall energy mix results in a higher penetration of renewable technologies. Payment by the demand is not modified by the inclusion of hybrid storage in the model formulation, since the substation remains the marginal unit.

Despite the fact that all the characteristics of the different storage techniques have not been applied in detail in the model, the potential of energy storage to integrate additional RES in the distribution network has been shown. Threats experienced by isolated systems as a consequence of the increasing RES penetration are higher than those experienced by interconnected systems, since they cannot depend on the smoothing effect of a large balancing area and interconnection flows. In this context, ESS may become a key element for the further integration of renewable energies.

Actual experience with integrating flexible DR and estimating own- and cross-price elasticities results insufficient to adequately evaluate its impact. It has been shown how DR can contribute to the deferment in capacity investment. Accurately determining the values for own- and cross-elasticity may significantly impact the outcomes of the planning process. Further research is necessary in both fields to adequately evaluate and incorporate its impact in the long-term distribution expansion planning problem.

1 SETS AND INDEXES

Y^l	Set of branches with feeders of type l
ν	Index for piecewise linear sections used for linearization of energy losses
$\Omega^l, \Omega^{LN}, \Omega^N, \Omega^p,$ Ω^{ss}, Ω^{st}	Sets of nodes connected to node i by a feeder l, load nodes, system nodes, candidate nodes for DG, substation, and storage nodes
i,j	Index for nodes
k	Index for alternatives for feeders, generators, and transformers
K^l	Set of available alternatives for feeders
K^{tr}	Set of available alternatives for transformers
l	Index for feeder types
L	Set of feeder types, $L = \{EFF, ERF, NRF, NAF\}$ where EFF, ERF, NRF, and NAF denote existing fixed feeders, existing replaceable feeders, new replacement feeders, and newly added feeders, respectively.
ll, lb	Index for load levels
LL, LB	Set of load levels
p	Index for generator types
P	Set of generator types, $P = \{W, \Theta\}$ where W and Θ denote wind and photovoltaic generators, respectively
st	Index for storage units
t	Index for time stages
T	Set of time stages
tr	Index for transformer types
TR	Set of transformer types, $TR = \{ET, NT\}$, where ET and NT denote existing transformers and newly added transformers, respectively
w	Index for scenarios

2 PARAMETERS

ε	Penetration limit for distributed generation
$\eta_i^{\text{st,prod}}$	Production efficiency rates for storage unit st
$\eta_i^{\text{st,store}}$	Storage efficiency rates for storage unit st
$\mu_{\text{ll},w}$	Loading factor of load level ll in scenario w
$\nu^l, \nu^{\text{NT}}, \nu^p, \nu^{\text{ss}}, \nu^{\text{st}}$	Lifetimes of feeders, new transformers, generators, and substation assets other than transformers
π_w	Weight of scenario w
$\xi_{\text{lb,ll}}$	Own- and cross-price elasticities for DR
A_ν	Upper limit for current flow corresponding to the piecewise linear section ν of the energy losses linearization
$A_{i,j,k,\nu}^l$	Upper limit for current flow corresponding to the piecewise linear section ν of the energy losses linearization through alternative k of feeder type l installed in branch $i-j$
$A_{i,k,\nu}^{\text{tr}}$	Upper limit for energy supplied corresponding to the piecewise linear section ν of the energy losses linearization for alternative k of transformer type tr installed in substation node i
$C_i^{\text{I,SS}}$	Investment cost of expanding existing substations by adding a new transformer or building a new substation
C_b	Costs for substation load level b
$C_{i,j,k}^{\text{I},l}$	Investment cost of alternative k of feeder l in branch $i-j$
$C_{i,j,k}^{\text{M},l}$	O&M cost of alternative k of feeder l in branch $i-j$
$C_i^{\text{M,st}}$	O&M cost of storage st at node i
$C_{i,k}^{\text{I,NT}}$	Investment cost of adding alternative k of a new transformer in substation node i
$C_i^{\text{I},p}$	Investment cost of generator p at node i
$C_i^{\text{I,st}}$	Investment cost of storage unit st at node i
$C_{i,k}^{\text{M},p}$	O&M cost of alternative k of generator p at node i

$C_{i,k}^{M,tr}$	O&M cost of alternative k of transformer tr installed in substation node i
$C^{E,p}$	Generation costs of generator type p
$\overline{C_{t,ll,w}^{SS}}$	Average substation cost at stage t for load level ll in scenario w considering adjacent blocks
$C^{st,prod}$	Production cost of storage unit st
$C^{st,store}$	Storage cost of storage unit st
$D_{i,t,ll,w}$	Expected demand at node i at stage t for load level ll in scenario w
$\tilde{D}_{i,t}$	Fictitious nodal demand in substation node$i$ at stage t
$\overline{F}_{i,j,k}^{l}$	Maximum capacity in branch $i-j$ at stage t for alternative k
$\overline{G}_{i,k}^{tr}$	Upper limit for energy supplied by alternative k of transformer type tr in substation node i
$\overline{G}^{p}$	Maximum capacity of generator type p
$\overline{G}^{st}$	Maximum capacity of storage unit type st
I	Annual investment rate
IB_{t}	Investment limit at stage t
$l_{i,j}$	Feeder length
L_{b}^{min}	Lower limit of substation load level b
L_{b}^{max}	Upper limit of substation load level b
n_{DG}, n_{T}	Number of candidate nodes for distributed generation, number of periods
pf	System power factor
$RR^{l}, RR^{NT}, RR^{p}, RR^{SS}, RR^{st}$	Capital recovery rates for investment in feeders, new transformers, generators, substations, and storage units
$\underline{s}_{i}^{st,store}$	Maximum storage capacity by storage unit st installed in node i
$\overline{s}_{i}^{st,store}$	Minimum storage capacity by storage unit st installed in node i
$\underline{s}_{i}^{st,prod}$	Minimum production capacity by storage unit st installed in node i
$\overline{s}_{i}^{st,prod}$	Maximum production capacity by storage unit st installed in node i

3 VARIABLES

$\delta^l_{i,j,k,t,\text{ll},\nu,w}$	Current flow corresponding to piecewise linear section ν of energy losses linearization through alternative k of feeder type l installed in branch $i-j$ at stage t for load level ll in scenario w
$\delta^{\text{tr}}_{i,k,t,\text{ll},\nu,w}$	Energy supplied corresponding to piecewise linear section ν of energy losses linearization by alternative k of transformer type tr installed in substation node i at stage t for load level ll in scenario w
δ_ν	Current flow corresponding to piecewise linear section ν of energy losses linearization
$v_{i,t,\text{ll},w}$	Voltage at node i, stage t, load level ll in scenario w
$v^{\text{st,prod}}_{i,t,\text{ll},w}$	Variables related to production at nodes i, stage t, for load level ll in scenario w
$v^{\text{st,store}}_{i,t,\text{ll},w}$	Variables related to storage at nodes i, stage t, for load level ll in scenario w
$\rho_{t,b,\text{ll},w}$	Variable associated with substation demand at stage t for substation load level b at load level ll and scenario w
$\theta_{t,b,\text{ll},w}$	Variable associated with substation demand at stage t for substation load level b load level ll and scenario w
c^{I}_t	Investment cost at stage t
c^{E}_t	Energy cost at stage t
c^{M}_t	Maintenance cost at stage t
c^{U}_t	Unserved energy cost at stage t
c^{ST}_t	Energy cost for storage units at stage t
$c^{\text{SS}}_{t,\text{ll},w}$	Substation price at stage t for load level ll and scenario w
$d_{i,t,\text{ll},w}$	Demand at node i at stage t for load level ll and scenario w when considering DR
$d^{\text{U}}_{i,t,\text{ll},w}$	Unserved energy at node i at stage t for load level ll at scenario w
$f^l_{i,j,k,t,\text{ll},w}$	Current flow through alternative k feeder type l installed in branch $i-j$ at stage t, load level ll and scenario w, measured at node i, which is greater than 0 if node i is the supplier and 0, otherwise

$\tilde{f}^{l}_{i,j,k,t}$ Fictitious current flow through alternative k of feeder l installed in branch $i-j$ at stage t measured at node i. Greater than 0 if node i is the supplier and 0, otherwise

$g^{p}_{i,t,ll,w}$ Energy supplied by generator p in node i at stage t for load level ll in scenario w

$g^{tr}_{i,k,t,ll,w}$ Energy supplied by alternative k of transformer tr installed in substation node$i$ at stage t for load level ll in scenario w

$\tilde{g}^{SS}_{i,t}$ Energy supplied by a fictitious substation at node i at stage t

$p_{t,ll,w}$ Demand payment at stage t for load level ll in scenario w

$s^{st,prod}_{i,t,ll,w}$ Power production for storage unit st at node i, stage t for load level ll in scenario w

$s^{st,store}_{i,t,ll,w}$ Power stored for storage unit st at node i, stage t for load level ll in scenario w

$x^{l}_{i,j,k,t}$ Binary variable for the installation of alternative k of feeder l installed in branch $i-j$ at stage t

$x^{NT}_{i,k,t}$ Binary variable for the installation of alternative k of new transformers in substation node i at stage t

$x^{p}_{i,t}$ Binary variable for the installation of generator type p at nodes i at stage t

$x^{SS}_{i,t}$ Binary variable for the expansion of existing substations by adding a new transformer or building a new substation in substation node i at stage t

$x^{st}_{i,t}$ Binary variable associated with the installation of storage unit st installed at nodes i at stage t

$\hat{x}^{st}_{i,t,ll,w}$ Storage level at nodes$i$ at stage t at load level ll and scenario w

$y^{l}_{i,j,k,t}, y^{p}_{i,j,k,t}, y^{tr}_{i,j,k,t}$ Binary utilization variables for feeders, generators, and transformers through alternative k installed in branch $i-j$ at stage t

$z_{t,b,ll,w}$ Binary variable associated with substation price at stage t for substation load level b, load level ll, and scenario w

$z^{dr}_{t,ll,w}$ Binary variable associated with DR at stage t for load level ll and scenario w

REFERENCES

[1] A. Gómez-Expósito, A.J. Conejo, C. Cañizares, Electric Energy Systems. Analysis and Operation, CRC Press, Boca Raton, FL, 2009.

[2] W.H. Kersting, Distribution System Modeling and Analysis, CRC Press, Boca Raton, FL, 2002.

[3] R.C. Lotero, J. Contreras, Distribution system planning with reliability, IEEE Trans. Power Deliv. 26 (4) (October 2011) 2552−2562.

[4] W. El-Khattam, K. Bhattacharya, Y.G. Hegazy, M.M.A. Salama, Optimal investment planning for distributed generation in a competitive electricity market, IEEE Trans. Power Syst. 19 (3) (August 2004) 1674−1684.

[5] Y. Baghzouz, Some general rules for distributed generation-feeder interaction, in: Proc. of the IEEE Power Engineering Society General Meeting, Montreal, 18−22 June 2006, p. 4.

[6] R. Viral, D.K. Khatod, Optimal planning of distributed generation systems in distribution system: a review, Renew. Sus. Energ. Rev. 16 (7) (September 2012) 5146−5165.

[7] G. Muñoz-Delgado, J. Contreras, J.M. Arroyo, Joint expansion planning of distributed generation and distribution networks, IEEE Trans. Power Syst. 30 (2014) 2579−2590. Available from: http://dx.doi.org/10.1109/TPWRS.2014.2364960.

[8] S. Haffner, L.F.A. Pereira, L.A. Pereira, L.S. Barreto, Multistage model for distribution expansion planning with distributed generation − part I: problem formulation, IEEE Trans. Power Deliv. 23 (2) (April 2008) 915−923.

[9] M. Stadler, C. Marnay, A. Siddiqui, J. Lai, B. Coffey, H. Aki, Effect of Heat and Electricity Storage and Reliability on Microgrid Viability: A Study of Commercial Buildings in California and New York States, LBNL, Berkeley, CA, 2009. Available from: <http://eetd.lbl.gov/sites/all/files/publications/lbnl-1334e.pdf>.

[10] P. Lombardi, Multi criteria optimization of an autonomous virtual power plant with high degree of renewable energy sources, in: Proc. of the 17th PSCC, Stockholm, 22−26 August 2011. ISBN 978-91-7501-257-5.

[11] EPRI-DOE Handbook of Energy Storage for Transmission and Distribution Applications, EPRI, Palo Alto, CA, and the U.S. Department of Energy, Washington, DC: 2003.

[12] Z. Styczynski, P. Lombardi, et al., Electric energy storage systems. Electra 255, April 2011, CIGRE Paris. ISBN: 978-2-85873-147-3.

[13] T. Gönen, B.L. Foote, Distribution-system planning using mixed-integer programming, IEEE Proc. C—Gener. Transm. Dis. 128 (2) (March 1981) 70−79.

[14] N. Ponnavaikko, K.S. Rao, S.S. Venkata, Distribution system planning through a quadratic mixed integer programming approach, IEEE Trans. Power Deliv. 2 (4) (October 1987) 1157−1163.

[15] Y. Tang, Power distribution system planning with reliability modeling and optimization, IEEE Trans. Power Syst. 11 (1) (February 1996) 181−189.

[16] W. El-Khattam, Y.G. Hegazy, M.M.A. Salama, An integrated distributed generation optimization model for distribution system planning, IEEE Trans. Power Syst. 20 (2) (May 2005) 1158−1165.

[17] H. Falaghi, M.R. Haghifam, ACO based algorithm for distributed generation sources allocation and sizing in distribution systems, in: Proc. of the 2007 IEEE PowerTech, Lausanne, 1–5 July 2007, pp. 555–560.

[18] S. Haffner, L.F.A. Pereira, L.A. Pereira, L.S. Barreto, Multistage model for distribution expansion planning with distributed generation—part II: numerical results, IEEE Trans. Power Deliv. 23 (2) (April 2008) 924–929.

[19] Y. Atwa, E. El-Saadany, M. Salama, R. Seethapathy, Optimal renewable resources mix for distribution system energy loss minimization, IEEE Trans. Power Syst. 25 (1) (Feb. 2010) 360–370.

[20] M. Lavorato, M.J. Rider, A.V. García, R. Romero, A constructive heuristic algorithm for distribution system planning, IEEE Trans. Power Syst. 25 (3) (August 2010) 1734–1742.

[21] K. Zou, A.P. Agalgaonkar, K.M. Muttaqi, S. Perera, Multi-objective optimisation for distribution system planning with renewable energy resources, in: Proc. of the IEEE International Energy Conference and Exhibition (EnergyCon), 18–22 December 2010, pp. 670–675.

[22] V.V. Thang, D.Q. Thong, B.Q. Khanh, A new model applied to the planning of distribution systems for competitive electricity markets, in: Proc. of the Electric Utility Deregulation and Restructuring and Power Technologies (DRPT), Weihai, 6–9 July 2011, pp. 631–638.

[23] A.M. Cossi, L.G.W. da Silva, R.A.R. Lázaro, J.R.S. Mantovani, Primary power distribution systems planning taking into account reliability, operation and expansion costs, IET Gener. Transm. Dis. 3 (2012) 274–284.

[24] M. Lavorato, J.F. Franco, M.J. Rider, R. Romero, Imposing radiality constraints in distribution system optimization problems, IEEE Trans. Power Syst. 27 (1) (February 2012) 172–180.

[25] E. Naderi, H. Seifi, M.S. Sepasian, A dynamic approach for distribution system planning considering distributed generation, IEEE Trans. Power Deliv. 17 (2) (July 2012) 646–653.

[26] X. Xiao, Z. Lin, F. Wen, J. Huang, Multiple-criteria decision-making of distribution system planning considering distributed generation, in: Proc. of the International Conference on Sustainable Power Generation and Supply (SUPERGEN 2012), Hangzhou, 8–9 September 2012, pp. 152–157.

[27] L. Hirth, Nota di Lavoro 90 The Optimal Share of Variable Renewables. How the Variability of Wind and Solar Power Affects Their Welfare-Optimal Deployment, Fondazione Eni Enrico Mattei, Milan, Italy, 2013.

[28] I. Ziari, G. Ledwich, A. Ghosh, G. Platt, Integrated distribution systems planning to improve reliability under load growth, IEEE Trans. Power Syst. 27 (2) (April 2012) 757–765.

[29] M.F. Shaaban, Y.M. Atwa, E.F. El-Saadany, DG allocation for benefit maximization in distribution networks, IEEE Trans. Power Syst. 28 (2) (May 2013) 639–649.

[30] D. Kammen, SWITCH optimization model, Lawrence Berkeley National Laboratory (online). Available from: <http://rael.berkeley.edu/old_drupal/switch>.

[31] M. Haller, S. Ludig, N. Bauer, Bridging the scales: a conceptual model for coordinated expansion of renewable power generation, transmission and storage, Renew. Sus. Energ. Rev. 16 (5) (March 2012) 2687–2695.

[32] A. Satchwell, R. Hledik, DE-AC02-05CH11231 Analytical Frameworks to Incorporate Demand Response in Long-term Resource Planning, Lawrence Berkeley National Laboratory, Berkeley, CA, September 2013.

[33] C. De Jonghe, B.F. Hobbs, R. Belmans, Optimal generation mix with short-term demand response and wind penetration, IEEE Trans. Power Syst. 27 (2) (May 2012) 830–839.

[34] A.K. Kazerooni, J. Mutale, Network investment planning for high penetration of wind energy under demand response program, in Proc. of the 11th International Conference on Probabilistic Methods Applied to Power Systems (PMAPS), 14–17 June 2010, pp. 238–243.

[35] M. Samadi, M.H. Javidi, M.S. Ghazizadeh, Modeling the effects of demand response on generation expansion planning in restructured power systems, J. Zhejiang Univ. Sci. C 14 (12) (December 2013) 966–976.

[36] S. Wogrin, P. Dueñas, A. Delgadillo, J. Reneses, A new approach to model load levels in electric power systems with high renewable penetration, IEEE Trans. Power Syst. 29 (5) (September 2014) 2210–2218.

[37] E. Lorenzo, G. Araujo, A. Cuevas, M. Egido, J. Miñano, R. Zilles, Solar electricity. Engineering of photovoltaic systems, PROGENSA, 1994, p. 316.

[38] Y. Atwa, E. El-Saadany, M. Salama, R. Seethapathy, Optimal renewable resources mix for distribution system energy loss minimization, IEEE Trans. Power Syst. 25 (1) (February 2010) 360–370.

[39] B. Borowy, Z.M. Salameh, Methodology for optimally sizing the combination of a battery bank and PV array in a wind/PV hybrid system, IEEE Trans. Energ. Convers. 11 (2) (June 1996) 367–375.

[40] L. Baringo, A.J. Conejo, Transmission and wind power investment, IEEE Trans. Power Syst. 27 (2) (May 2012) 885–893.

[41] L. Baringo, A.J. Conejo, Correlated wind-power production and electric load scenarios for investment decisions, Appl. Energ. 101 (January 2013) 475–482.

[42] S.J. Kazempour, A.J. Conejo, C. Ruiz, Strategic generation investment using a complementarity approach, IEEE Trans. Power Syst. 26 (2) (May 2011) 940–948.

[43] G. Strbac, E.D. Farmer, Framework for the incorporation of demand-side in a competitive electricity market, IEEE Proc. Gener. Transm. Distrib. 143 (3) (1996) 3–8.

[44] P. Palensky, D. Dietrich, Demand side management: demand response, intelligent energy systems, and smart loads, IEEE Trans. Ind. Informat. 7 (3) (2011) 381–388.

[45] C. Gellings, The concept of demand-side management for electric utilities, Proc. IEEE 73 (1985) 1468–1470.

[46] Eurelectric, EU Islands: Towards a Sustainable Energy Future, June 2012.

[47] S.D. Braithwait, K. Eakin, L. R. C. Associates, The Role of Demand Response in Electric Power, Edison Electric Institute, Washington, DC, October 2002.

[48] K. Spees, L.B. Lave, Current market, Electricity J. 20 (3) (2007) 69–85.

[49] K. Dietrich, J. M. Latorre, L. Olmos, A. Ramos, Demand Response and Its Sensitivity to Participation Rates and Elasticities, in Proc. of the 8th International Conference on the European Energy Market (EEM), Zagreb, Croatia, 25–27 May 2011, pp. 717–716.

[50] A.K. Kazerooni, J. Mutale, Transmission Network Planning under a price-based demand response program, in: 2010 IEEE PES Transmission and Distribution Conference and Exposition, 14–17 April 2010, New Orleans, LA.

[51] K. Dietrich, J.M. Latorre, L. Olmos, A. Ramos, Demand Response in an Isolated System with High Wind Integration, IEEE Trans. Power Syst. 27 (1) (2012) 20–29.

[52] F.C. Schweppe, M.C. Caramanis, R.D. Tabors, R.E. Bohn, Spot Pricing of Electricity, Kluwer Academic Publishers, Norwell, MA, 1988.

[53] EWEA, Large scale integration of wind energy in the European power supply: analysis, issues and recommendations. A report by EWEA. EWEA, December 2005.

[54] B.F. Hobbs, H.B. Rouse, D.T. Hoog, Measuring the economic value of demand-side and supply resources in integrated resource planning models, IEEE Trans. Power Syst. 8 (3) (August 1993) 979–987.

[55] L. Gkatzikis, I. Koutsopoulos, T. Salodinis, The role of aggregators in smart grid demand, IEEE J. Sel. Areas Commun. 31 (2013) 1247–1257.

[56] IRENA, Electricity Storage and Renewables for Island Power, IRENA, 2012.

[57] B.R. Alamri, A.R. Alamri, Technical review of energy storage technologies when integrated with intermittent renewable energy, in Proc. of the International Conference on Sustainable Power Generation and Supply, SUPERGEN '09, 2009.

[58] J. Eyer, G. Corey, Energy storage for the electricity grid: benefits and market potential assessment guide, in: S. REPORT (Ed.), Sandia National Laboratories Albuquerque, New Mexico 87185 and Livermore, California, 2010, pp. 1–232.

[59] S.O. Geurin, A.K. Barnes, J.C. Balda, Smart grid applications of selected energy storage technologies, in: Proc. of the IEEE Innovative Smart Grid Technologies (ISGT), 2012.

[60] D. Pozo, J. Contreras, Unit commitment with ideal and generic energy storage units, IEEE Trans. Power Syst. 29 (6) (November 2014) 2974–2998.

[61] L.T. Blank, A.J. Tarquin, S. Iverson, Engineering Economy, McGraw-Hill, New York, NY, 2005.

[62] A.L. Motto, F.D. Galiana, A.J. Conejo, J.M. Arroyo, Network-constrained multiperiod auction for a pool-based electricity market, IEEE Trans. Power Syst. 17 (3) (August 2002) 646–653.

[63] GAMS—The Solver Manuals, GAMS Development Corporation, Washington DC, December 2012.

[64] Electricity demand prices for Lanzarote-Fuerteventura, Red Eléctrica de España, 2014 (online). Available from: <http://www.esios.ree.es/web-publica/>.

[65] R.E. Rosenthal, GAMS—A User's Guide, GAMS Development Corporation, Washington DC, December 2012.

Printed in the United States
By Bookmasters